I0816133

# PRAISE FOR *THE NATURE OF FASHION*

'From the Heart of the World, the Sierra Nevada de Santa Marta, the Kogui people thank Carry Somers for her important mission, which is intimately connected with our ancestral culture. More than her human self, we thank her great spirit for recognising her mission here in the physical world and for helping to spin, weave and remember the colourful consciousness of our still-living ancestry.'

**MaMo Sintana** and **MaMo Isuama Kunchaluintana**, Kogui spiritual leaders

'When I look at a piece of clothing, I see the seeds of its origins – the plants it was made from and the soil they were grown in. In *The Nature of Fashion*, Carry has dived deep into the origins of these textiles, telling us stories of craft and creativity as well as politics and exploitation, weaving the threads of fashion's past into uplifting hope for its future. More than a history of fashion, this is a fascinating unravelling of how fabric has shaped us and the world we live in.'

**Arizona Muse**, founder of DIRT, Earth activist and model

'*The Nature of Fashion* is a lyrical reckoning of the roots of what we wear. Carry Somers traces the lineage of plant-based fibres with the precision of a researcher and the soul of a storyteller – unearthing histories often left out of fashion's mainstream narrative. What results is both a return and a roadmap: a vision of fashion that is not extractive but ecological, cultural and deeply alive.'

**Aditi Mayer**, climate storyteller, sustainable fashion advocate and founder of The Artisan Archive

'As I watch the trees and plants grow around me, I see how we are intricately connected to nature. Carry captures these threads and weaves an absorbing story of how we first fashioned clothes from plants and how we might once again work closely with the natural world. Along the way, Carry crosses centuries and continents, revealing hidden stories of craft, creativity, science and innovation as well as trade and politics. An unmissable book.'

**Adam Clayton**, U2

'Masterfully weaving together history, anthropology and mythology, Carry bridges the realm between the ancient and modern, helping us to reimagine our relationships with our clothes. On every page, she reminds us how the textiles that adorn our bodies have the potential to connect us back to the earth and our ancestors, bringing us back into harmony with the living world.'

**Nathalie Kelley Mallqui**, Quechua storyteller

'This book is a delight. It draws together a marvellous tale of human invention in search of materials to clothe and sustain us and a rollicking ride of experimentation, colonialism and piracy. Carry raises thrilling questions about how techniques and materials now nearly extinct might be reimagined in a new world where technology can give nature a helping hand. *The Nature of Fashion* is both history and future in the same breath. I closed the book wondering which pages contained the seeds of a new future or a new fortune – I loved it.'

**Sir Tim Smit**, co-founder of
The Lost Gardens of Heligan and the Eden Project

'Carry Somers' poetic and politically engaged prose is infused with a love and deep respect for Indigenous customs and centuries-old woman craft. Underpinned by both scholarly and historical research and chronologically mapped out from 41,000 BCE to the current day, Carry's knowledge and storytelling empowers and spellbinds us. I am deeply moved.'

**Caryn Franklin MBE**, fashion and identity commentator

'Reading *The Nature of Fashion* reminded me of *A History of the World in 100 Objects* – the way a single material can open a window into centuries of human story. Carry Somers weaves history, the natural world and personal journey into a powerful reminder that what we wear connects us to the earth – or separates us from it. An urgent meditation on our material world, and a thoughtful guide for those seeking meaning in fashion.'

**Nishanth Chopra**, founder of Ōshadi Collective

'Carry has woven a beautiful narrative about threads and the nature that provided them, from the very beginning of humankind. Her global stories illustrate how we are destroying our planet but also give us hope that nature will one day prevail, if we can learn the lessons it teaches us. Fashion is what clothes us and protects us, but it should not be something that leaves such a toxic trail behind us.'

**Jane Shepherdson CBE**, chair of My Wardrobe HQ

'*The Nature of Fashion* is a fascinating read, a real eye-opener. Now, every time I get dressed, I'm reminded of another story of where our clothes originated, dating back thousands of years. It's superbly researched, yet Carry Somers writes so beautifully, filling in the gaps and conjuring up the scenes for us. The history and politics of textiles has been turned into quite a page-turner!'

**Joe Swift**, garden designer, writer and broadcaster

'A tapestry of sciences from anthropology to archaeology, botany to biology, carpology to climate change. *The Nature of Fashion* is more than an understanding of sustainable fashion practices, it is a guidebook that postulates how fabric strung our history together and how it is directing our species' future.'

**Dr Gabby Wild**, National Geographic author, wildlife veterinarian and conservationist

'A fascinating read that delves into the roots of materials and the way they've been treasured by communities as well as exploited by colonisers around the world. People and plants have been interwoven since the beginning of time, but these relationships have been incredibly difficult to maintain as powerful forces have made it their mission to tear them apart, with dire consequences. Carry writes succinctly about this history while also highlighting the ways we can all turn the narrative around for the better.'

**Amelia Windsor**, fashion writer and creator

# THE NATURE OF FASHION

# THE NATURE OF FASHION

## A BOTANICAL STORY OF OUR MATERIAL LIVES

CARRY SOMERS

Chelsea Green Publishing
White River Junction, Vermont
London, UK

First published in 2025 by Chelsea Green Publishing | PO Box 4529 | White River Junction, VT 05001 | West Wing, Somerset House, Strand | London, WC2R 1LA, UK | www.chelseagreen.com
A Division of Rizzoli International Publications, Inc. | 49 West 27th Street | New York, NY 10001 | www.rizzoliusa.com



Publisher: Charles Miers
Deputy Publisher: Matthew Derr
Commissioning Editor: Muna Reyal
Project Manager: Susan Pegg
Copy Editor: Susan Pegg
Proofreader: Jacqui Lewis
Indexer: Charmian Parkin
Designer: Jenna Richardson

ISBN 978-1-915294-79-1 (hardcover) | ISBN 978-1-915294-80-7 (ebook) | ISBN 978-1-915294-81-4 (audiobook)
Library of Congress Control Number: 2025024268 (print)

**Our Commitment to Green Publishing**
Chelsea Green sees publishing as a tool for cultural change and ecological stewardship. We strive to align our book manufacturing practices with our editorial mission and to reduce the impact of our business enterprise in the environment. We print our books on chlorine-free recycled paper, using vegetable-based inks whenever possible. This book may cost slightly more because it was printed on paper that contains recycled fiber, and we hope you'll agree that it's worth it. *The Nature of Fashion* was printed on paper supplied by Sheridan that is made of recycled materials and other controlled sources.

Authorized EU representative for product safety and compliance
Mondadori Libri S.p.A. | www.mondadori.it
via Gian Battista Vico 42 | Milan, Italy 20123

Printed in the United States of America.
10 9 8 7 6 5 4 3 2 1    25 26 27 28 29

*For my parents, Robert and Lesley,*
*who planted the seeds of curiosity.*

*For Mark, the warp to my weft.*

*And for Sienna, my wildflower.*

# CONTENTS

# PREFACE

This book is rooted in rigorous research, yet moves with freedom, filling in the gaps when archaeology, archives and oral histories fall silent. It is, in part, a work of imagination, particularly for the earliest stories where evidence is patchy at best, while striving to remain true to the lived experience of those times. Inspired by Eduardo Galeano's *Memory of Fire* trilogy, each episode is marked by its year and place, with countries referred to by their present-day names. Direct transcriptions from primary sources, such as books, journals and manuscripts, have been woven into the text in italics, preserving the original voices of the past. For Indigenous Peoples, I have endeavoured to include both their self-designations or earlier names and those by which they are known today.

Where sources conflict, I have sought to prioritise firsthand accounts or oral traditions where possible. The silenced are not in the archives – their stories have been suppressed, erased or never written. Absence speaks as loudly as the words that remain, many of which are recorded through the lens of colonists, explorers, anthropologists, priests, whose perspectives are coloured by their times and intent. To fill in the spaces, I have attempted to evoke the texture and nuance of lived history rather than impose a static narrative.

The inequality of the archives, especially in matters of gender, class and race, became increasingly evident the deeper I burrowed into this research. The record-keepers determined whose lives would be preserved for future generations, and their silence is most deafening when it comes to those they deemed insignificant; the stories of ordinary women are largely missing from the chronicles, which were designed to showcase the feats of extraordinary men. Women remain at the selvedge of history – stabilising the seams, preventing the fabric from fraying – yet their contributions are too often overlooked, their voices hovering at the margins while men fill in the words. When the written narrative excludes one half of the story, a margin of doubt runs down the edge of every page. The same is true for Indigenous Peoples around the world, then and now.

History is an imperfect record, and so, too, is any attempt to recount it. The responsibility for how these stories are told is mine alone. This book does not claim neutrality; history rarely is. Every choice – what to include, what to omit – carries weight. My selection of stories is shaped both by personal interest and by what I believe will make a compelling narrative. I consciously chose to omit many well-worn story strands in order to make space for the unknown and overlooked. Throughout my research and writing, my aim has been to listen to what the voices of the past have to say, allowing them to speak in their own words where possible, and to imagine with sensitivity where records fail them. This is their story, woven from history's imperfect threads.

# INTRODUCTION
# A Single Thread

What if the story of humanity could be told through a single thread? Long before we inked our thoughts onto parchment, perhaps even before we etched stories onto stone, we began spinning plant fibres. Textiles are among our earliest attempts to fashion the world around us, and their history is inseparable from our own. The Latin verb *texere*, meaning to weave or construct with elaborate care, reminds us that both textiles and texts stem from acts of creation. Textiles and texts, born of the same etymological thread, are entwined with our histories, our cultures, our lives. From the very first moment someone twisted and plied plants into thread, textiles became a means of communication, an expression of the soul of the community where they were born. Threads, like words, are woven together: strands twisted into stories, each with its unique texture. In the beginning was the cloth.

It is not only in Latin that the two are united. The *Popol Vuh*, the Maya *Book of the Community* written in the K'iche' language, opens with the words: *This is the root of the Ancient Word, here in this place called K'iche'. Here we shall write, we shall plant the ancient stories of the beginning.*[1] Planting was traditionally carried out with a stick that pierced the ground, opening up a space for the seeds in the same way as a needle pierces cloth for thread to pass through, so some communities naturally started using the same word for the intricate embellishment of their textiles. Thus, an alternative translation might read: *This is the root of the Ancient Word, here in this place called K'iche'. Here we shall create, we shall brocade the ancient stories of the beginning.*[2] Through planting with digging sticks, the maize is born. When they sit down to weave, the cloth is born. And when communities gather to spin wondrous tales, conjuring images out of glimmering threads, a new narrative comes to life. Even the name *Popol Vuh* evokes the connection between text and the material world: *Popol* signifies a woven mat, symbolic of the indivisible community, while *Vuh* refers to a book. The Dogon of Mali share a similar understanding. They use the word *soy* for cloth. It means the spoken word too. Across the globe, every culture has its way of interweaving cloth with community, the underlying fabric of our existence. Textiles have always been more than just material.

Foundational to societies around the world, textiles reflect the choices we make and the traces we leave behind. It was this understanding of textiles as a mirror of human activity that carried me on a 2,000-mile sailing journey from the Galápagos Islands to Easter Island, one of the most remote inhabited places on Earth. My mission was to uncover the fate of our clothes on their invisible journeys beyond our wardrobes. Early seafarers navigating that route would have studied the sea's shape and colour to determine their proximity to land. On our scientific voyage, we were likewise scanning the ocean for meaning, but we were searching for traces of distant hands. Yet as the wind filled our sails, propelling us towards our lonely landfall, we saw no signs of human life – no boats, no plastic, not even a bottle. Just the glittering phosphorescence of a billion nighttime plankton, pods of dolphins and whales, and the occasional flop of a flying fish hitting the deck. Still, we knew this wasn't the full story. For hours every day, we sifted the Pacific waters, seeking clues to the hidden world beneath. Under our microscopes, the ocean revealed its secrets: a translucent filament coiled inside my sample bottle, shimmering like a star; a black mesh; flecks of blue, orange, green and white. Multicoloured clouds drifting in a sea of blue. These bright fragments, remnants of textiles and other materials, were but small indicators of the vast trail of plastic waste left in the wake of uncontrolled growth.

We arrived on Easter Island as the wildest of storms unleashed its fury, sending garden furniture pirouetting over rooftops, blowing out the power supply and sinking boats in the harbour. The following morning, I made my way across the island to the iconic Moai statues on Anakena beach. But it was the sand beneath my feet that caught my attention: I stood mesmerised by a kaleidoscope of colour. This was where all the plastic was ending up, breaking down into smaller and smaller pieces. Cleaning it up felt impossible, like trying to hold back the tide with my hands. I turned my gaze to the Moai, ancient statues believed to possess spiritual powers for the benefit of humanity. How I wished they would use their potency to clean up this mosaic of human neglect and disregard. When I turned to leave, it was more than the wind stinging my eyes. Yet in the midst of my despair, I saw hope. If the Moai weren't coming to our rescue, perhaps we could save ourselves. For I believe that within us all lies the power to mend what is broken and to protect our planet from the ravages of our own making.

I returned to England in March 2020, just as the country was shutting down; I wasn't going anywhere for a while. Anchored to my table – a landlocked sailor in a lockdown world – I began sifting through my memories of the voyage, separating out fragmented thoughts and scrutinising them for meaning. Where did those fibres originate, and what harmful chemicals might they be carrying? How could I help stop the contamination of our seas?

I also started to wonder about the roots of today's ecological crisis, how far back into our history, philosophy and culture it extended. Earlier civilisations made mistakes as well; the barren slopes of Easter Island, once lushly forested, were stark evidence of that, just as microplastics were confirmation of our own. How far would the lifeless tendrils of our plastic present extend? I pictured a future Mary Anning chipping away at a technofossil trapped in artificial permafrost. As the long-preserved relic emerges, it turns out to be a faux-fur leopard-print coat, its sorrowful form embodying a far-distant era. I started wondering about other fibres too, such as those made from trees. Did they melt back into the soil from which they once sprouted, vanishing like snow with the onset of spring, or would they, like plastic, be with us for an eternity? With a shake of my head, I tried to dispel these unsettling images from my mind. Surely, a more promising future lay ahead? It was simply a matter of finding a star to steer by.

And so, I set out to discover what brought us here, how we can move forwards, and how to avoid repeating old patterns. As we stride forwards in progress, we sometimes forget to turn our heads and glance back, and it's difficult to grasp the full extent of our loss without understanding how things were in the past. Charting an unmapped course, I began navigating through time to a world beyond the confines of my present – a vast new ocean of unseen dangers and boundless possibility. I crossed the globe and ventured back to the dawn of humanity's life on Earth. And I'd like to take you there too. Because in order to understand where we are, and where we're going, we need to know where we've come from.

All journeys begin at home, and over the course of this book we will tack back and forth between past and present. The places may be familiar, the cast of characters less so. Yet the stories of yesterday will find their echoes in today, because these are, at their heart, human stories. We will trace the origins of fibres and dyes and their rich interconnection with the earth as revealed through myths, symbols and worldviews handed down over generations. We will glimpse the communal spirit embedded in early textiles, silent witnesses to countless hours of devotion. As we open ourselves to the wisdom of our ancestors, we may even hear Earth's quiet murmurs beneath our feet, although there are times she sighs too deeply for words. Continuing our journey, we'll see how textiles transformed human and natural realms, uncovering tales of passion, blindness, idealism, obsession and the age-old struggle for riches and power. As the centuries unfold, we will learn how we often struggled to make the right decisions, failing to recognise the signals of nature. How we made mistakes – often devastating ones – and how we learnt from them. Taken in isolation, these are just stories. But together they represent so much more. Every culture, every place, every plant is part of an ecosystem, every thread part of a pattern, revealing how the choices we once made shape the landscapes and economies around us today. Each strand contributes to the text that will weave the finished textile, just as the clothes we are wearing embody hundreds of entangled stories.

Today, there is a growing movement, a revival of sorts, that seeks to restore our connections to Earth, reclaiming ancient wisdom and creating textiles in harmony with nature, not working against it. This book is part of that movement. It is a call to action, a reminder that the choices we make today will shape the fabric of our world for generations to come. Ultimately, this is more than a history of how we transformed plants into cloth: it is a map to our material future, one where plants are again revered as the very foundation of life. As they rightfully are. In a curious twist of etymology, the word map derives from the Latin *mappa* meaning a signal cloth or flag and, just as a flag marks direction, the unfurling story of textiles offers clues to the paths we might choose tomorrow.

With that in mind, let's travel back to where the story of textiles begins, more than forty thousand years ago. To a single thread.

# PART ONE

# In the Beginning Was the Cloth

## String

*France • 41,000 BCE*

What do you see when you look at a piece of cloth: a mesh of interwoven strands or tendrils of possibility? Either way, the creation begins with a length of string – a thread that binds, yet is boundless, spinning into patterns that have long shaped our world.

Imagine living during the Ice Age, a time when Earth's climate was in constant flux. Temperatures could change fast, swinging from biting cold to glacial in a matter of decades. In this unforgiving environment, Neanderthals, our close evolutionary cousins, were engaged in a constant struggle for survival. Clad in simple hides and furs that offered little protection from the elements, they were forced to migrate south as temperatures plummeted, seeking refuge in more temperate climes. As advancing ice pushed them to the fringes of their known world, they clung on to life in the caves of Gibraltar, seemingly with nowhere else to go. Failure to adapt to the changing climate is a long-standing theory around their demise.

Then archaeologists uncovered something extraordinary. At Abri du Maras, a rock shelter on the west bank of the Rhône in modern-day France, they found a small piece of three-ply string, crafted from the satin-soft inner bark of a conifer and tied around a stone tool. It dates back to between forty-one thousand and fifty-two thousand years ago.[1] String may seem like a simple artefact, but those flimsy fibres challenge long-held assumptions about Neanderthals' capabilities. These people were not just survivalists but resourceful and creative beings who were endeavouring to adapt to the changing world around them, like their contemporaries the early *Homo sapiens*.

As my thoughts fly through the ages to distant frozen times, I wonder what guided the string's crafting? Was this innovation born out of necessity, or reflective of a more sophisticated culture we are yet to fully comprehend? In an effort to understand the faint strands of our past, I try to picture how early string was made.

---

String is not a simple thing. It ties, it binds, it fastens, it holds. And whatever they do with it, string must be strong, for these are tough people – they have to be when the temperatures are glacial and the best way to get meat is to stab a reindeer up close with a spear. Broken bones and fractured skulls are only to be expected in their perilous world. Every journey is a risk, a reminder of how precarious life can be.

Within the dense pine forest above their spring campsite, the air is alive with the cries of white-tailed eagles and the haunting bellow of giant deer, while the understory rustles with spirits, benevolent and otherwise. Frosted boughs hang low with the weight of the snow. A woman crunches a pine cone underfoot. She glances around warily, scanning the shadows before continuing on her way. Eventually, she finds what she is looking for and, flint in hand, begins stripping the deeply furrowed bark from a black pine, knowing the most supple fibres will be found now when the sap starts to rise; wait too long and they'll turn thick and woody. All the while, her ears are alert to the rustling of branches, her eyes darting from side to side watching for bears. There are no such trees in the lowlands where her family spend their winters, so she must make the most of this opportunity, gathering all she can carry.

Squinting, she emerges from the deep jade of the forest, despite the dull lid on the sky. Although they had been hoping for a warmer year, winter still has a hold on the land: the sky is leaden, and conifers creak with the weight of snow. Spring has yet to fully arrive. Below, on the alluvial plain, spotted horses run wild and, as she pauses for a moment to watch their careening, she notices how the river has narrowed between its two muddy flanks. Less water is never a good sign. Clasping the bark tight to her breast, she pushes on, scrambling down the frost-patterned slopes. All the while, the freezing wind whips her face.

Back at the family's rocky shelter, snowflakes are swirling again. The woman lays the bark pieces on the ground and begins beating them with a dull stone to strip away the soft inner layer – the part that she needs. Then, she takes these ribboned strands to soak in a nearby pond to loosen the gummy fibres, hoping muskrats won't steal them for their twiggy domed homes. She only learnt the technique a year or two ago, and it's taking some time to master, but one thing is certain: if she wants fine thread, she cannot hurry this stage, no matter how eager she is to make string. A few days later, she breaks the ice on the pond with a heavy stone and pulls the fibres from the water, hanging them to dry in the whipping wind. Then, she begins rolling the soft strands against her thigh, slowly growing the threads.

All around her, the camp is a bustle of activity – children gathering firewood, women preparing food – but the next step will be hard for her to do alone. She calls her older daughter to help, and together they work, one holding the end of the fibres while the other twists, making the string. And as they work, the woman spins wondrous tales about the beginning of time, the legendary feats of the ancients, the antics of the wood spirits, the time her father wrestled a cave lion, so their past will not be forgotten. As the afternoon stretches on, the pile of string grows. Once all the fibres have been spun, the woman selects three lengths and hands one end of each piece to her daughter. Taking up the other ends, she twists the threads together in the opposite direction, following the passage of the moon through the sky. This, she has found, will make her string strong: hard to unravel, less likely to break.

With such a thing as three-twisted string, she can make just about anything. Perhaps she will coil it into a snare to trap a wild boar? Then, with the rest of it, she could fasten a wooden handle onto a sharp flint and use the tool to skin the creature too. Or maybe she will loop it into a bag for gathering roots of wild carrot, parsnip and cattail, or coil a basket for storing earthnuts, because it's never too early to start preparations for the cold months to come. If she can make her string tough enough, it could even bind logs into a raft that could float down the slow-flowing river – and if they reached the sea, she could knot a net to throw into the waves and haul great catfish ashore. She could, but she won't, for the woman has plans for this string: to fashion something other than the draped furs and skins her family have worn for a lifetime.

To make their usual attire, she squats over a woolly mammoth carcass, cutting away its warm skin with a flint or shell, scraping and tanning it quickly, softening the skin using the creature's urine and brains in a process all too familiar to her. Once the hide is prepared, she trims it to shape, then pierces holes along the edges with a deer bone awl. In and out she goes, threading a strip of leather, using the awl to push it through the holes, working methodically down the seam until the two sides are bound together. It's an improvement on what they used to wear, but at the end of the day it's still a few lengths of animal skin draped over the body. That won't stop the merciless wind drying out their insides. Her family needs better if they are to survive this changing climate, like the clothes she saw on those tall people, their faces unfamiliar, their gait long and light.

When she first sighted them over the ridge, she had been astonished; their garments were as strange as wild pears putting forth their spring blossoms in an entirely different colour. It was as if the world had transformed without her noticing. An inkling in her belly told her this clothing helped them survive the relentless cold. If only she could figure out how to make such garments herself, using twists and knots to tailor their furs and skins, filling in the gaps where the weather bores in. Maybe then they wouldn't lose so many of their children? It's worth a try, at least. The men, of course, will think she is wasting her time. A cloak of skin over the shoulders has served their purposes since time immemorial. And as for her own children, the woman can hear the echo of her daughter's wails in her head, her face scrunched up in displeasure at the thought of cutting up her cherished aurochs-hide cape and sewing it back together in a more figure-hugging form.

As the woman squats on the ground, she lays out the string in front of her, running her fingers over its serpentine form, examining its strength. Perhaps she should decorate the cords, staining them with the same yellow ochre they use to paint their bodies, weaving in eagle talons, cave lion teeth or carved earthen beads. Maybe then her daughter will wear it? She will put all her effort into creating something beautiful. But the woman understands this is more than ornamentation. More even than a new form of expression. If they want to survive this fluctuating climate, this is the way of the future. She feels it deep in the marrow of her chilly bones: it's time for her people to embrace wearing plants too.

---

Whoever made that single thread displayed a profound understanding of their environment. Perhaps, they observed how the outer bark of a conifer offered the tree protection from moisture loss and insulated against harsh weather, yet over time they realised it was the fibrous inner bark, made of living cells, that was essential for the tree's growth and future survival. Furs and hides played the role of outer bark, yet they needed more. Did the Neanderthals set out to experiment, to imitate, learning from the layering ways of plants to increase their chances of survival? This ingenuity speaks volumes about their intrinsic connection with the natural world and their ability to utilise the resources to hand. But I believe this string does more than provide an insight into their inventive spirit. It serves as a reminder to look to nature for inspiration as we fashion our future within a fast-changing world.

## A lifeline

*Georgia • 34,000 BCE*

Eight thousand years have passed since the family group of Neanderthals occupied their rock shelter – a mere blink of an eye in the age of the world. Yet over this short period of time, these close relatives of modern humans have vanished. As the climate changed and food supplies became uncertain, they simply couldn't adjust fast enough to the shifting world around them. The ability to adapt, including the adoption of new fibre technologies, allowed early modern humans to flourish in a glacial world. Though *Homo sapiens* still made their clothing from furs and hides, it was cut and shaped to fit the body, perhaps layered for protection against the bone-chilling wind. Crafting such garments required specialised tools: blades for cutting skins, awls for piercing holes and needles for sewing the pieces together. And, of course, the women – for archaeologists have attributed this role to women, although it may not have been exclusively so – required string or thread, which they had to learn how to harvest, process, spin, ply, braid and sew. It is clear that string was no humble invention. Indeed, it was likely a crucial factor in the survival of early humans, enabling us to spread out and inhabit almost every biome on our planet.

String can be made from many things. In the foothills of the Caucasus Mountains in present-day Georgia, a region that in ancient times shared the same frigid temperatures and relentless winds as much of Europe, archaeologists uncovered some of the earliest known examples of processed wild flax. These insubstantial remains weave an illuminating story, revealing the vital role played by plant fibres in the flourishing of early human societies. Although adopting fibrous finery wasn't without some small challenges.

---

A northwesterly wind howls through the mountains, and its breath bears a glacial chill. Inside the rock shelter, the firelight flickers, illuminating the handprints and animals the clan have painted with ochre on the rough limestone walls, stories of where they have come from and dreams of where they will go. In the glow of the firelight, a young woman is working, head bowed over shimmering strands. Moistening one end of the thread with her lips, she squints, moves towards the flames, and passes it through the eye of a bone needle. Now she can start to sew. Her arm dances, rising and falling; the movement is intuitive, as if it has been ingrained in her brain since the beginnings of time, in the same way she knows how to gather nuts, roots and seeds. Yet, she is one of the few who understands the secret of the tall plants down by the river. Delicate plants with thin stems, narrow leaves and small purplish-blue flowers, you would never suspect them of yielding such fibres, supple and strong, with the honeyed glow of the sun. Not long ago, her clan had relied solely on animal hides for warmth, but the changing climate demanded adaptation, and that required innovation, even if some are less keen.

Across the fire, her mother – an elderly woman approaching thirty winters in age – watches icily. The young woman lets out an involuntary shudder and pulls her wolverine cape up to her chin, not from the blast of wind sweeping through the cave's entrance, but the weight of her mother's chilling stare. She may yet force her daughter to sacrifice her hard-won threads, having made no secret of her thoughts on the futility of this endeavour. The young woman has to admit the moths are a nuisance. The beetles too. No sooner has she gathered, soaked, dried, combed and spun the wild flax into thread than the pesky insects come along and start munching the fibres. But she isn't giving up now, not for the sake of a hole or two. The world around her is changing, the ice sheets pursuing their creeping advance. As the clan moves from place to place, they find refuge in caves such as this to escape the biting winds that press in from the north, shivering their skin and freezing their bones. Yet even with fire and shelter, the cold demands more.

At night, when the clan gather around the fire, the elders often recount stories from far-distant days when Earth was warm, forests were lush and game was plentiful. Now, the animals are mostly gone, moving on to new territories in search of plants that know how to survive this terrible cold. Her brother has seen reindeer and musk oxen prints in fresh-fallen snow; the animals are moving north to the tundra. Of course, she knows what their legends say about the mountains having legs so they can walk from the edge of the ocean to the desert, but in all the years they've been coming to this cave, those great rocky domes haven't moved a hair's breadth. No one is going to save them but themselves, and without following the animals, they will surely starve. She only hopes to find such fibres wherever they go.

And there's something else, although she scarcely dares admit it. The colours! Who wants to be draped in beastly brown hues when these threads can be dyed pink with autumn bilberries, green with nettles, yellow with sour sorrel, purple with knotweed, black with oregano? The land around their cave may be rugged and sparse, but it holds hidden treasures for those with the imagination to explore. This golden fibre is more than a tool: it holds the essence of the earth, the plants and the minerals, connecting her to the land she has called home these past five winters. If the clan must move with the game, these radiant strands will live on as an afterglow of where they have been – a way to animate their past like the stories told by the elders.

At the entrance to the cave, the crunch of snow makes her tense. Is it man or beast? She quietly reaches for her spear, but it's no more than the heavy tread of returning hunters. No words are spoken, none needed, because they find their echo in the empty rumblings of her belly. The woman glances across at her baby daughter who has drifted off to sleep on a bearskin by the fire. In her tight little hand, she grips a small twist of flax. Could it be that the knowledge of this gilded plant will ensure her survival? These fibres are more than string; they are a lifeline.

---

The significant role of string in the story of humanity raises questions about whether archaeology may have overlooked the importance of textiles. Why isn't there a Fibre Age in our historical divisions of time? The imbalance in archaeological findings is perhaps not surprising given the perishable nature of fibres, unlike more durable materials such as stone, antler and bone, which receive more attention. However, in rare environments like caves and burial sites where textiles have been preserved, they are often found in abundance, outnumbering less perishable artefacts. Could this lack of emphasis stem from the fact that textile production was primarily domestic, performed by women in pre-market societies?

Simple string played a powerful role in taming the world to human will, enabling the creation of snares, fishing lines, tethers, leashes, nets, handles and ties. It provided a way to securely bind hard materials together, enabling the construction of more complex tools. In her book *Women's Work: The First 20,000 Years*, Elizabeth Wayland Barber dubs it the String Revolution.[2] While there may not be an era named after it, the spinning of thread and the invention of string rank among humanity's most pivotal advancements, on a par with the creation of stone tools and the discovery of fire. Yet early *Homo sapiens* were not solely focused on survival: they sought beauty as well. The fibres found in Dzudzuana Cave in the Caucasus foothills were more than utilitarian, their multiple colourways suggesting both an aesthetic appreciation and a cultural expression. Beauty, as William Morris observed, isn't merely incidental to our lives, but a positive necessity.[3]

Just as our early ancestors in Dzudzuana Cave recognised, embracing plant fibres in our wardrobes does not mean sacrificing colour – far from it! I spend more than half the year living in a caravan in the Undercliff in Branscombe, a liminal world caught between sea and sky after the cliff slid away. The view from my window reminds me of the interplay between nature and time, each season, storm and shift in the landscape leaving its mark. For years, I worked to untangle the fashion industry's problems, but here, surrounded by flora, the answers seem disarmingly straightforward: pendulous sedge and nettles for fibre; madder, bird cherry, elder, alkanet or weld for colour.

Yet scaling up from garden to garment is no easy task. And herein lies the real challenge: finding ways to draw on practices like these without losing sight of what innovation can add. Or what the plants themselves might be able to teach us if we'd just pause to listen. As I reflect on how ancient civilisations adapted to the changing climate, I'm confident we, too, can discover new ways to coexist with nature as we face the challenges ahead. If we do, I believe the fabric of tomorrow can be more resplendent – and yes, more inherently beautiful – than anything we have known.

## A river in the desert

***Israel* • *8200 BCE***

As the Ice Age lost its grip, early humans began to spread across the land, venturing into areas abandoned by the retreating glaciers. Caribou, muskoxen, mammoths, lions, bears and other creatures roamed freely through ice-free corridors, crossing the Bering Land Bridge between Siberia and Alaska. A vast expanse of former ocean floor, sagebushes and sphagnum moss now sprouted there instead of seaweed. Where animals went, humans weren't far behind. Some early Eurasian peoples made daring voyages down the coast of the Americas during this period of dispersal, but any campsites are now deep beneath the waters. Thousands more were already living on the continent, having journeyed across the land bridge thousands of years earlier, in days when Earth enjoyed a more welcoming climate. Excavations near the Yana River in Siberia have unearthed evidence of human activity from over thirty thousand years ago. Bone awls, eyed needles and needle cases, alongside carbonised fabric remains demonstrate early humans' ability to craft essential items, such as tents, underwear, outerwear and accessories. But they didn't stop there – the presence of beads and fine needles reveals both a love of personal adornment and a means of cultural expression. These well-prepared and skilled people likely journeyed eastward to the Americas, adapting to new environments along the way. As time passed and nature shifted once more, any ideas of a return journey would have become impossible.

Whenever and however they travelled, fibre production was one essential skill to carry on their journey into a new hemisphere. Upon arrival, they would have found a wealth of unfamiliar plants waiting to be studied, experimented with and turned into string. Remains discovered in Paisley Caves in present-day Oregon dating back fourteen thousand years show evidence of basketry and cordage made from dogbane, milkweed, daisies and stinging nettles. Expanding wetlands provided ideal conditions for plants like rushes, sedges, cane and cattail to flourish while the dry steppe was rich with sagebrush, bitter brush, cliffrose, sumac and milkweed. Decoration could take the form of wolf lichen dye, bird feathers or the equally feathery white bear grass. The possibilities for transformation were endless.

As early settlers in the Americas were harnessing the power of plants to adapt to their changing environments, a parallel story of human ingenuity was unfolding halfway around the world. In the arid landscape of the Judean Desert, ancient peoples domesticated wild flax and transformed it into cloth. Then they went further, developing advanced techniques for preserving and decorating their sacred items. When these relics were uncovered, they revealed not only the earliest preserved linen fabrics in the Old World, but the first examples of plant-based coatings. Intrigued, I decide to take a closer look.

---

The small cave overlooking the river can only be reached with difficulty; no one wants hyenas and leopards picking over the founding generation's sacred skulls. Nor can they risk any creature destroying the beautiful objects they hide there. Coated with a glue made from animal skins and a fragrant tree gum, these treasures should last an eternity in the dry desert air. Over the years, they add more and more to their ritual hoard. There's a life-sized limestone mask with red and green stripes, and carved stone heads no bigger than their thumbnail. Baskets hold cattle bones, sea squill and other objects for use in their ceremonies; made from rushes and reeds and coiled into cords, their designs alternate layers of darkness and light. Everything is protected from the elements by the same odoriferous glue.

Then there are textiles from pale, wiry-stemmed, fine-leaved flax, a plant that looks so fragile they find it hard to believe it can survive their arid soil and rocky ground. But thrive it does, almost reaching their waist, flowering blue or lilac, even white. And despite its short-lived blooms, opening at daybreak and fading with dusk, the fibres drawn from its stems will last a lifetime. After years of gathering the plant in the wild, one of the reasons they settled down by the river on this dry asphalt plain, some distance from the desert cave where their men perform sacred ceremonies, was to cultivate flax. Growing their own makes a difference. Unlike the straggle of plants in the wild, when they sow the seeds they can pack the plants tight together; with support all around, they grow taller, making their fibres longer. The seeds go in early, after the first inundations of winter are past; by the time the lunar month of Adar comes around, the valley will ripple with flowers. One hundred days from their sowing, the stalks will have yellowed and all but one or two of the flowers will have dropped. That's when the crop is ready to be picked: pulled up by the roots and laid out to dry, seeds stripped away with a coarse comb.

For as long as anyone can remember, they have pressed precious oil from these seeds, but now the women press its fibres into service too – for ropes, shrouds and even the occasional garment. But first there's the soaking, the rinsing, the drying, the beating and yet more combing to be done. It all takes time, but eventually they have soft strands of flax to roll against their thighs, overlapping the ends, lengthening the fibres, before twisting the strands onto wooden spindles with carved limestone whorls. To give the flax even greater strength, they might ply it with itself, or with rushes or reeds from the shallow margins of the river. They will even use hair combed from their heads, twisting two or three strands together until the tension is just right.

Now, it is time to decide what to create. Sometimes they make nets, coaxing the thread through a needle's eye, twisting a loop, then another, until they line up in a row. Circling, twisting, knotting, the loops spread outward, downward, forming a network. This might turn into a headdress, decorated with lime plaster beads or shells from the distant sea, traded for dried figs and olive oil. But here in the Judean Desert, between the cliffs of Mount Sodom and the shore of the Dead Sea, women prefer to use their skills to create more elaborate cloth; after all, they have been weaving since the dawn of time. Well, perhaps not quite since the dawning if the stories are true of how the world was once cold, the earth frozen and their ancestors wore nothing but furs, not like these days when it has grown so hot and humid that wearing anything other than flax would be unimaginable.

One of the women is weaving, two weft threads taut between her fingers. One goes under the warp, one over. She twists them together and does this again. One under, one over, twist and repeat. Backwards and forwards she goes, time after time, building up layers, creating the cloth. Then she prepares the glue, simmering bones from their domestic cattle to extract their collagen, adding grated root of sea squill to prevent damage and decay. While the fragrant resin tapped from sweetgum trees was closer to hand, this was reserved for coating their sacred skulls. Once the gummy mixture has cooled, she spreads it over the textile with deliberate strokes, sealing in protection. Now her cloth is ready to withstand a lifetime in the desert.

---

The shift from foraging to farming was part of broader economic and social changes during the Neolithic period, ushering in new social structures and a range of technological innovations, including a surprising variety of burial customs. While the cloth found in Nahal Hemar Cave was made from flax, the baskets were twined from the strands of reeds, rushes and grasses, which were coated inside and out, most likely for waterproofing. When archaeologists tested the baskets, one sample revealed high amounts of a protein found in the bulbs of sea squill, a Mediterranean wildflower that blooms at summer's end. But it was a resin tapped from the small, sweet-scented styrax tree and from species of liquidambar – the word amber in Arabic means an aromatic liquid – that was reserved for coating the decorated skulls.

These resins were perhaps a surprising choice given the widespread use of natural bitumen, a substance found in oil sands and pitch lakes of petroleum. Bitumen has been used throughout human history – Neanderthals even used it for gluing handles to their tools. During the Neolithic era, it served as a common protective coating for everything from baskets and burial clothing to boats. Bitumen's ubiquity raises the question of why these people didn't use the fossil fuels readily to hand given that they lived on an asphalt plain, sitting atop a rich source of bitumen they seemingly chose not to exploit. Perhaps they were unaware of its existence, experts speculate. But isn't it also possible that here, in one of the cradles of civilisation, sacred skulls were simply too precious to be daubed with a sticky tar-like substance derived from long-decomposed matter? Surely, only the purest vanilla-scented gum resin was considered worthy of honouring the dead?

Bitumen continues to shape our modern world. Derived from the distillation of crude oil, it lies beneath our feet as we walk down the footpath, beneath our car wheels as we travel along the road, coating the bridge we cross to prevent corrosion. When we return home, it might be there too, shielding our roof against the elements. While bitumen is no longer used to protect textiles, as it was in the past, modern waterproofing formulations come with a dangerous drawback: many contain PFAS, known as 'forever chemicals' due to their accumulation in wildlife and human bodies. Yet amid this hazardous landscape, there are alternatives if we choose to seek them out. Plant-based coatings from renewable materials like food industry by-products not only offer sustainable solutions but can match the performance of their synthetic counterparts. Silk treated with tea-stem waste displays durable flame-retardant properties, lotus leaf extract enhances the water repellency of dyed polyester yarns, while grape seed extract is reported to provide cotton fabric with excellent UV protection, along with antibacterial, antioxidant, flame retardant, anti-static and anti-pilling benefits.[4] This being so, should we not then adopt the same ethos as those who ornamented these ancient skulls, shunning fossil fuels for materials that uphold the sacredness of all life on Earth?

## The cow and the oak

***Turkey • 6400 BCE***

On the vast Southern Anatolian Plateau, the ancient settlement of Çatalhöyük has been hailed as an early protocity, a precursor to modern urban life. Thanks to the cultivation of cereals and the domestication of livestock, people were able to settle here year-round, the tightly packed dwellings fostering a bustling community with intricate social structures and rich cultural practices. Textile technologies were integral, as evidenced by the production of plain weave textiles. Along with the site of Jarmo in northern Iraq, it is among the earliest known examples of this craft. As the settlement thrived, the need for resources and space became increasingly pressing. I'd like to take you back to Çatalhöyük at a time when everything changed for the residents of that city.

---

There are eight thousand men and women living in Çatalhöyük. Not long ago, all were equal, men and women alike, their society bound not by leaders, but by shared convictions, ceremonies and magnificent artwork. The growth of the city was hardly surprising given its favourable position straddling two hills, one to the east, one to the west, with the Çarşamba River running between. Hackberry, almond, plum and pistachio all grew on the steppe, with stands of oak here and there. Further west rose forests of juniper and yet more oak, while a lake lay to the east. Between the wetlands on the valley floor and the dry higher ground, this land provided all that they needed and more, allowing them plenty of time to practise their artistry.

In fact, art became something of an obsession in Çatalhöyük; its citizens harnessed the riches of the earth – minerals such as red and yellow ochre, cinnabar, azulite, malachite; some scraped from slender rock veins, others obtained through trade – to leave their mark on this Earth. Walls, ceilings, even the furniture of their mud-brick homes were plastered in white, lime-based clay. Then, grinding up rocks with a pestle and mortar to make pigments, they would add decorative flourishes, sometimes mixing in obsidian to deepen the paint's lustre. Most of the time they painted birds and animals, although not their domestic sheep and goats that grazed on the plain – they could see those by climbing the ladder and poking their heads out of the doorway in their ceiling, the houses being so tightly crammed together that rooftops served as streets. No, the people of Çatalhöyük wanted to depict their mastery over the natural world so future generations would remember them as a brave and fearless people. And so, they painted themselves dressed in leopard skins, baiting wild bulls, stags, boars, raptors, a bear. If they happened to kill a wild creature, they inserted its tusks, jaws and claws into the plasterwork. Here is proof, they would say, that the people of Çatalhöyük are stronger than the world around us.

But most of the time they didn't dress in leopard skins, because there weren't eight thousand leopards to go around. Instead, they wore something equally majestic and far more accessible: the oak. They chopped down the trees in their thousands, using the wood to support the walls of their homes, for roof timbers, as fuel, to fire pottery, sculpt vessels – so why not use it for their clothes? It was far easier than importing flax from over the mountains. Scraping away the bast between the bark and the wood, they would lay the fibres end to end, overlapping them just a little before splicing and rolling into one long thread. The fabric they made was uneven, porous and rough to the touch, but it smelled like a woodland, sweet with sap. Oaks were everywhere; so many, in fact, no one imagined a day when that yawning expanse of trees would be emptied of green.

So much has changed. And it's not only the trees. Now, some are more equal than others. It never used to be that way, not so long ago, when everyone shared the work, no one accumulated wealth and everyone made art. This has brought tension to Çatalhöyük where before there was none. Fights break out, people firing clay balls with their slings and hitting one another over the head. But that's the least of the problems. The cows are probably to blame. The neighbours over the mountains had them, so the people of Çatalhöyük wanted them too. At least some did. But the cows needed water and grassland, so they burned the forest, still leaving some stands of oak to build houses and make clothes. But some people got greedy, always wanting more. These days there are tens of thousands of cattle all over the plain, not to mention the sheep and goats. To grow grass for all those cows, they irrigated the land by the river, back before the river dried up. Now the landscape resembles a desert, its soil blown like dust in the wind. One or two oaks remain, their leaves yellow-tipped, curling up, dropping off. A few skeletal cows still stalk the ravished ground, lowing like lost creatures from some bestial underworld. Others are less fortunate. Entire ribcages pierce the rim of red earth, like the collapsed frame of an abandoned storehouse, while other carcasses lie white and gleaming as wave-turned shells on a distant sandy shore. The land is little more than a backdrop for death.

No one realised the oaks offered more than wood for construction and fibre for cloth. Not until they were gone. Only now do the men and women of Çatalhöyük understand how their canopy provided shade, protecting the ground from hot, drying winds. How fallen leaves enriched the fecund humus, locking moisture underground. How their roots anchored the riverbank, protecting against erosion. How everything was connected: the cows, the oaks, the permanence of their city. Everyone thought Çatalhöyük would be there for ever. No one believes that anymore.

---

Reflecting on this ancient city and its fate has led me to reconsider some of my earlier assumptions. Before I started researching this book, I had placed the separation of humans and nature squarely on the shoulders of Enlightenment thinkers like Isaac Newton, René Descartes and Francis Bacon. These men, instrumental in shaping the modern Western worldview, effectively sanctioned our right to exploit nature in the name of material progress by diminishing all non-human entities to instruments of purpose. Now, I discover the rift between humans and the natural world began thousands of years earlier. Driven by a desire for expansion, combined with an emerging appetite for consumer goods, the people of Çatalhöyük pushed deeper into the wilderness, clearing the land, encroaching on the surrounding forest, felling the trees.

And those cows opened a sluice gate of unintended consequences. It's little wonder the site's lead archaeologist described Çatalhöyük as an early example of how the advent of farming ended our symbiotic relationship with nature.[5]

But is that the full story, or did the people of Çatalhöyük learn from the cascade of cause and effect that devastated their ecosystem? Did they start anew elsewhere, carrying hard-won wisdom with them? Perhaps community elders began weaving stories, embedding cautionary tales into their cultural fabric so future generations would avoid repeating the same follies. The true legacy of Çatalhöyük may lie not just in its remarkable ruins, but in the lessons its people sought to pass down, resounding through the ages to reach us today.

## Strange fruit

***Papua New Guinea • 4350 BCE***

It turns out that the people of Çatalhöyük were not alone in clearing their land of trees. Far from the dry plains of Anatolia, a different kind of transformation was unfurling in the Western Highlands of Papua New Guinea.

---

In the misty dawn of the highlands, where the land teases the sky and the sky cries incessant tears, lies the swampy embrace of Kuk. When people first arrived in this wetland paradise, it was carpeted with the sort of plants that love these moist soils: tree ferns curling their tendrils over a mosaic of spiky grasses and sedges, with tufts of reeds here and there. Away in the distance, lush forests closed in, cloaking sheer slopes, their peaks lost to the clouds. It was at the fertile margins of these wetlands that the first stirrings of agriculture took root, tethering the destiny of a nomadic people to a soggy patch of earth. But it was up in the highlands, in the heart of the forest, where they first came across the precious tree. Among the tallest of understory plants, all competing for light beneath the dense canopy, its fruit gleamed a full spectrum of colour: orange, green, red, blue – and, yes, yellow too. Feathered arms open wide, the tree looked to be singing, offering itself to the people. They heeded its call. Back to the damp margins they brought it, burning patches of forest to make space for this jewel-strung plant. On the edge of the swamp, they began transforming banana into gold.

Thousands of years in the future, the tree will be prized for its sweet-tasting fruit – seeds bred out, plants rendered sterile – yet in these early times, its crescent-moon fruiting is barely edible, a mouthful of grit. No, it is not for its bananas they honour this tree, but the myriad other gifts it bestows. Male flowers are cooked, as are stems, leaves and tubers. When people fall sick, the roasted fruit stops diarrhoea, young flower buds bring relief to the breathless and the outer stem treats insect bites. Leaves become wrapping, vessels, covers, plates, a mat for weary bodies, a shelter and raincoat for long hunts and foraging journeys. The stems provide fibres that are twined into cordage and knotted into nets to harness the bounty of the rivers.

And then there's clothing, for while it's nowhere near as frigid as the thinly remembered past, it can still be mighty cold in the highlands. And when it rains, how it rains! With weather like this, skirts and capes are essential, yet such garments are far from merely functional. For the people of Kuk have found banana sap will dye a rich brown, its flower petals produce soft yellows, while ash from the stem fixes colours onto the cloth. Yes, the people of Kuk have learnt that if they stay still long enough to listen to the earth, they can coax out its gifts with persuasive hands. Indeed, so adept do they become at this new agriculture, they clear more of the forest to grow more and more banana, as well as taro and yam.

When the understory grows back, they burn it again, planting crops across the valley floor and lower slopes where forests once stood tall. Now, only the highest slopes retain their verdant mantle.

As the years flow on, the lowland soil grows tired, its nutrients drained, but venturing further up the mountains holds no appeal for these people of the valley. There is only one way to go. They turn their eyes in the direction of the swamp. Digging channels, pits and runnels, they drain water from the marsh to tap into the fertile promise enclosed within the damp earth. The insects mock their efforts – the island teems with more creeping, buzzing life than anywhere else on the planet – yet still they persevere, their work accompanied by a flapping of faces and slapping at legs. Expanding into this wet ecosystem will protect their bananas from drought, providing a lifeline to a future they have yet to imagine.

Seven thousand years later, archaeologists will uncover the remains of this place, this swamp called Kuk, and recognise it as one of the world's earliest cradles of agriculture. And they would say that the people settled here to cultivate wild plants for food. But was it only to supply their menu? Could it not be material matters – textiles and tools – that were just as vital to the flourishing of this early civilisation? After all, it has happened before.

---

As we will witness time and again, deforestation for clothing and textiles has deep historic origins. Yet in traditional societies like those of Papua New Guinea, respect for nature is fundamental to their way of seeing. With time and experience, they would come to understand the importance of balance and reciprocity, learning not to take too much or simply gather what was close at hand. They would see how careful harvesting helped the forest to flourish, reducing competition and allowing energy-giving light to flood in. As wisdom grew, they wove it into their stories. But stories need to be continually spoken, because dead words are no help to any living soul.

Today, we have become disconnected from the symbiotic relationship our ancestors once shared with the natural world. Distanced from the consequences, it becomes all too easy to dismiss the global impacts of our choices. According to Canopy, an organisation dedicated to conserving forest ecosystems, around three hundred million trees are logged every year to produce man-made cellulosic fabrics like viscose, often from our most ancient and endangered woodlands. That's enough trees to encircle Earth seven times. Even trees from responsibly managed forests or fast-growing species like bamboo and eucalyptus are not the simple sustainable solution they might appear because the transformation from tree to textile comes with its own set of challenges.

In a process euphemistically termed regeneration, wood pulp is dissolved in chemicals to create a thick, soupy mixture, which is forced through tiny holes in a spinneret, like a showerhead, to form long, thin fibres. These are hardened using chemicals, washed with more chemicals, spun into thread and dyed with yet more chemicals. Wood liquefied, spun and reborn, this transformation severs trees from nature in ways we are only just beginning to understand; now that their cellular structure has shifted to a form not found in nature, there is little that knows how to break it down. Yet the problem extends beyond regenerated fibres. The broader category of cellulosics – encompassing everything from abaca, the contemporary name for banana fibre, through to cotton – are no less implicated, accounting for over seventy per cent of fibres collected from freshwater, marine waters, animals and the atmosphere.

But what if we could rebuild our relationship with nature? What if we truly understood how deeply our lives are entwined with the world's forests? These questions inspired me to set off on this journey: to explore radical visions of what textiles could be like if made in more circular, cooperative, regenerative ways. What if natural really meant natural, disappearing into the earth to nourish the fibres of tomorrow? The word radical comes from the Latin *radicalis* meaning something with roots, fundamental to the natural processes of life. Radical is about going deep, to the root of the matter, as well as spreading out through living systems, interweaving, becoming part of an interconnected web, and I began to wonder if it wasn't time to sift through the debris and unearth fashion's roots, rediscovering the stories we've forgotten along the way. Because when we know where we've come from, it becomes easier to see where we are going. Renewal, after all, begins from the ground up.

## Weaving is knowledge

***Peru • 4200 BCE***

The Levant, Anatolia, Papua New Guinea… It's no coincidence that some of the oldest textile remnants emerge from the first cradles of agriculture. As people settled in one place, those early societies discovered the luxury of time: to observe nature's rhythms, to develop their skills, to experiment with plant fibres, ultimately turning them into cloth. With the emergence of complex social systems, textiles took on a crucial role: supporting economies, facilitating trade, upholding beauty, reinforcing beliefs, symbolising power. From birth to death and onwards into the afterlife, textiles wove their way into the fabric of civilisation.

This connection is perhaps most evident in Peru, another early heartland of agriculture. At Huaca Prieta on the country's northern coast, archaeologists unearthed textile fragments and twined fibres dating back eleven thousand years. Among the world's oldest surviving cotton textiles, they display complex, colourful designs, revealing early domesticated cotton in hues of red, white and brown, alongside the earliest-known example of indigo-dyed cloth. But these textiles were more than adornment: they were knowledge encoded. The repetition of patterns across various samples, the directional shifts in twining to create texture, and the presence of motifs like the condor all point to a sophisticated community-wide design system. The history of Huaca Prieta is not just one of archaeological discovery – it is a story spun from innovation and creativity, and the intrinsic understanding that to weave is to know.

---

Every story has to start somewhere, and usually there's an ending as well. But not in Huaca Prieta. Here, time moves in a spiral, looping back on itself every thousand years. Each new era is birthed by a *pachacuti*, a cataclysmic event in which the world heaves itself up and rebalances. In this place where the beginning endlessly begins again, the line between history and myth becomes blurred.

According to legend, the people of the first age knew nothing, not even how to make clothes. Although they could twist fibres, they still dressed in leaves. In the second age, people began to work, tilling the fields, building irrigation ditches, yet they still did not know how to make clothing, wearing animal skins instead. The people of the third age learnt to spin and weave. When they went into battle, it was cloth and clothing they wanted as the spoils of war. Everyone in Huaca Prieta is happy they are living in the third age.

In their fertile coastal home where the glacial meltwater of the Andes flows out to sea, the inhabitants of Huaca Prieta have more time than those living elsewhere. Because their society has no high-class rulers to serve and appease, it gives them ample opportunity to innovate and experiment, to develop technologies, craft tools, paint motifs and create exquisite cloth. They don't even need to fight battles anymore; why would they when they live in continual reciprocity with the world around them, trading the fruits of the sea for roots and grains from their neighbours in the mountains? The rich coastal silt provides perfect conditions for cultivating their native cotton, which grows like the rainbow in brown and orange, cream and yellow, green and mauve. Their cotton has always been naturally colourful, but now its shades are deeper, more muted; selective breeding has helped make nets and lines less visible to fish. And while they were modifying the plant, they lengthened its fibres to make it easier to spin. So, while the men go fishing with their brown cotton nets, dried gourds attached for floats, the women sit down to work in the shade of the huarango trees.

A woman is preparing tan-coloured cotton bolls for spinning, teasing out their fibres as if funnelling clouds. The light-as-dust threads will provide a pleasing contrast to the indigo steeping in her dye vat. Yet she is not content with just the one fibre – this is Huaca Prieta after all, a place where conventions are challenged and boundaries pushed. She selects several lengths of wild chivo vine and plies it with the indigo-dyed cotton. Then, tying her backstrap loom to a huarango trunk, she makes herself comfortable on a mat on the ground. Using two weft yarns, she raises one over the warp and pushes the other one under; twist and repeat. As she works, she moves her body this way and that to create tension in the vertical threads. It's slow work, but she has all the time in the world.

What people do in life is reflected in death. Women are buried with their colourful textiles and looms, while their menfolk make do with fishing nets, pelican feathers and other paraphernalia from their maritime pursuits. These gendered grave goods join them in the dark pyramid at the centre of their world. Bonfires burn on its slopes through the night and glittering sparks rise up to join with the stars. People imagine life will go on as it always has: their sons will fish, their daughters will weave and everyone will thrive with maize from the mountains and fish from the sea.

So it has been, so it always will be. The people of Huaca Prieta cannot foresee the great pachacuti when time and their world will overturn once again.

In the far distant future, a descendant of their mountain neighbours will write a very long letter to the King of Spain, whose crown now spans these ancestral lands. His name is Felipe Guamán Poma de Ayala, born of the sky and the earth, the condor and the puma. In his epistle, he portrays the people of the third age as multiplying like sand on the seashore. The pure people, he calls them, illustrating his words with a drawing of a couple outside their hut, the woman holding a spindle, the man toying with a ball of yarn like a yo-yo. No, the people of Huaca Prieta cannot anticipate the chaos of conquest, but Guamán Poma sees it, believing the final cycle of the last sun has already begun. Turning the page, he depicts two armies preparing for battle: one, the Spanish conquerors, unleashing brutality and discord; the other, the Indigenous *awka pacha runa*, warriors of the earth, their resistance as fierce as the invaders advance. *Awka* means warrior, but it has another meaning too: someone who is unbending, unwilling to change their ways or face the consequences of their actions. In this collision of immovable forces, Guamán Poma sees chaos just over the horizon.

Yet there is hope, for a pachacuti is a liminal space, a transitional period in which the balance between human beings, earth beings and spirits is disrupted. A time flush with potential. If only people would tune in to planetary rhythms and listen to Pachamama's slow-beating heart, they could restore balance to this upside-down world. After all, there is no finality to an ending if it heralds a new beginning.

---

The resplendent work of Huaca Prieta's textile artists speaks of rules and rituals lost to the years, of knowledge encoded in cloth.

A blue-tinged woven fragment found at the archaeological site, believed to be around 6,200 years old, is no simple textile, reflecting the deep weaving tradition at the heart of Peruvian culture. Cotton has been domesticated in Peru for at least 8,000 years, and this piece shows evidence of generations of selective breeding for longer, stronger fibres. This same botanical understanding extended to other plants in their landscape as well. All around the world, early cultures discovered various indigo-producing plants and developed processes to extract and ferment the pigment, but the Huaca Prieta textile was dyed with the most prized of all: *Indigofera tinctoria*, true indigo. And whoever made it didn't stop there. The indigo-dyed cotton was plied with fibre spun from the chivo vine, a type of milkweed endemic to Peru, and the yarn was then woven across biscotti-coloured cotton warp threads. With its fine narrow stripes, the resulting textile would look at home in any resort collection today. The Huaca Prieta textile is not just an archaeological curiosity, but an important chapter in the rich history of making fabric from flora.

Writing in 1615, seventy years after the conquest of Peru, Felipe Guamán Poma de Ayala observed how neglecting *ayni* had plunged his era into turmoil. Ayni means balance, but it is no simple concept. It was ayni that set in motion the inflow and outflow of creative ideas, a continuous pairing and repairing that formed the essence of life. It was ayni that moved their community beyond self-interest, encouraging work for the common good, because no one and nothing could survive on its own. It was ayni that facilitated exchanges between highland and lowland dwellers, so everyone could thrive. It was ayni that opened up conversations between all forms of life, speaking and silent, visible and invisible, for in Guamán Poma's world plants do more than just grow and are valued far beyond mere utility. Ayni is the ongoing act of reciprocity that sustains the whole world, and a continual accompaniment to life. For regeneration to take place, it must be recreated anew every day in solidarity with the community of all living things.

Yet amid the tumult of his era, Guamán Poma saw a chance for renewal, believing that time binds the knowledge of the past with the spirit of the present. A pachacuti is more than an ending – it is an opportunity for a new beginning. In our hands lie the threads of all our pasts, and we can choose how to weave them into our futures. Yes, there have times when we have lost our way, when ignorance, laziness or greed intruded on our relationship with the natural world, creating a gnarly knot in the midst of a harmonious pattern. But when we got it right, it was glorious! And in every twist of every thread lies the possibility of getting it right once again.

## A hole in the earth

***Brazil • 2500 BCE***

Beyond where condors soared above glacial peaks lay an impenetrable jungle. The people of Peru's third age believed this to be the domain of their ancestors, living as they had in the past, still clad in leaves. While the inhabitants of Huaca Prieta were illuminated by knowledge, they remained in the dark about their distant neighbours, who had been spinning and weaving for thousands of years.

---

Deep in the forest, the Wuy jugu people thrive, growing arrowroot, yams, sweet potato, bottle gourds, squash, wild harvesting from forest palms, while their clear rivers teem with fish. The Wuy jugu express themselves through art too, painting rocks and caves with pigments drawn from natural minerals: iron oxide for red and yellow, carbon and manganese for black, kaolin for white, mixing them up with resins, eggs, animal fat and water. Having little need for clothing in their hot and humid climate, they pattern their bodies with fruits of the forest instead. Yet cotton is central to their lives. Indeed, it is why they are there.

According to tales passed down through the elders, when Karosakaybu, the creator, first shaped the world, humans had yet to be born. One day, his companion, Daiiru the armadillo, burrowed deep into the earth, where he discovered a world turned on its head: the sun rose in the west and journeyed eastward across the sky, while daytime on Earth was moonlit in the underworld. But it was not the topsy-turviness that astonished him most. In that hidden world, the armadillo found something else unexpected.

When Daiiru spoke of what he had seen, Karosakaybu wove a rope from the finest cotton, which he lowered into the depths, with the armadillo tied to one end. What the armadillo had found was indeed wondrous – men, women and children living underground – and since there were no people on the surface of Earth, Daiiru led them out. Up the cotton lifeline they climbed, but as more and more ascended, the weight became unbearable. With a crack like a lightning bolt, the rope snapped. Half the people tumbled back to the underworld where they remain to this day. To mark the survivors as his own people, Karosakaybu painted their skin with deep blue-black patterns, drawn from the unripe genipap fruit. Then, he taught them all they needed to thrive, including how to cultivate and spin the cotton that saved them, the thread that bridged their two worlds.

In their time, the Wuy jugu have mastered the art of transforming these diaphanous white clouds into strong, silken threads, and their craft has taken on new forms. They twist and knot cords into nets, capturing food from the rivers and sky, and hammocks for resting in at the end of the day. They scale the leafy canopy to retrieve sinuous creepers or muriti palm leaves. By the river, they trim rippling grasses, weaving them into baskets of all shapes and sizes. Their hands move methodically as if connected by a thread to their minds, which are filled with visions of ancient times, guiding their strokes, giving shape to their stories.

In years to come, others will call them by a name not their own: a rival tribe will label them the Munduruku, the red ants, for their tactic of fighting in large groups. But in their own language, they remain Wuy jugu: our own, our people – a name as entwined with their identity as the cotton they spin.

---

A few years ago, I travelled up the Amazon to visit rubber-tapper communities in the Tapajós National Park, where I saw how the Munduruku's sacred lands are under constant pressure from mining companies and agribusiness. Monoculture soy flanked the road, stretching as far as the horizon, replaced by thirsty cotton in soy's off-season. Meanwhile, the river created by Karosakaybu, once teeming with fish, is poisoned by mercury from mining and toxic agrochemicals from the cotton. The Munduruku's sacred spaces are vanishing before their eyes. To them, this land is so much more than a resource, more even than their home: it is a storied map of their existence, where place and time intertwine. Like a palimpsest of memory layering up the years, every bend in the river, every rise and fall in the terrain, is etched with histories stretching back to the dawn of time. Living across multiple physical and temporal realms, their struggle goes beyond defence of territory – it's a fight for the very ecology of life, a web of interconnection that underlies all existence.

Beyond the cotton in the Munduruku's origin story, another layer drew me in: the muriti palm, known as *aguaje* or *ité* across Latin America. At home in my caravan, naturally dyed, backstrap-loomed mats made from its fibres adorn the floor, crafted by the Urarina women of the Peruvian rainforest. While they serve as mats to sleep on at night, shrouds to envelop their dead and currency for trading with other groups, these weavings are more than just functional objects. They bind history to the present and weave it into their futures.

Thriving across the Amazon basin, the muriti palm is a giant of the forest, flourishing in dense clusters within *aguajales* – peat-rich wetlands dominated by the species that serve as almighty carbon sinks. Spanning millions of hectares, these ecosystems store three to five times more carbon per hectare than most other tropical forests, making them a critical buffer against climate change. Yet, they too are under threat from development, agriculture and mining. Even the fruit, an important source of nutrition for the local population, has become a point of vulnerability – some trees are felled for easier harvesting, a far simpler task than shimmying up a hundred feet to the forest canopy. Muriti palms are slow growing, and once destroyed this critical habitat will be almost impossible to restore.

But the Amazon's story has always been one of coexistence. Agriculture and hunting have shaped its landscapes for thousands of years, with local communities enriching biodiversity while stewarding the forest. Peruvian fashiontech company Neofibers, co-founded by my friend Sumy Kujon, exemplifies this approach. Partnering with local communities to harvest aguaje from the muriti palm, Neofibers transforms the leaves into fine thread while promoting agroforestry and implementing reforestation programmes. Such initiatives not only boost local economies but incentivise the preservation of these unparalleled ecosystems, furnishing us with sustainable fibres along the way. This partnership of tradition and innovation reflects a way forwards. By embracing practices rooted in respect, we can support both vibrant ecosystems and flourishing rural economies in a post-carbon world.

## Why accuracy is important

*Mexico • 1097 BCE*

While the majestic *aguaje* and soft pillows of cotton interwove the lives of the Wuy jugu, thousands of miles to the north, the Olmec revered an altogether squatter and spikier plant. But while *maguey* – Nahuatl for agave – may be a relatively low-lying species, it certainly punches above its weight when it comes to fibre. The tough leaves make cordage, baskets, textiles and sleeping mats known as *petates*, while some plants secrete a soapy wax that can be used to wash clothes. The Olmec found a multitude of uses for versatile maguey, including fermenting its sap into an alcoholic beverage they called pulque, served to priests to boost their enthusiasm for the rituals of sacrifice. Maguey's needle-sharp thorns were indispensable for such purposes as well.

---

A maguey thorn passed through the penis heightened the spiritual charge of bloodletting rituals, although a cord twisted from the plant's fibres and drawn through a hole in the tongue nurtured the ancestor gods too. These are some of the ceremonies that take place in their sacred cave where the Olmec climb to a higher realm of consciousness in order to carry out their water rites. Etched by centuries of wind and rain, they call this place *The Mountain Where the Flower of the Maguey Plant Grows*, and in case any future peoples should forget its name, the men grind haematite and paint the soot-blackened walls with a flower rising from a maguey plant atop a mountain.[6] Beneath their feet, the limestone floor is scattered with thorns and splattered with phallic blood.

The Olmec rulers don't need to go far to find maguey – the plant seems to thrive at any altitude and on any soil – but co-opting its power requires a whole ritual toolkit. It would affront the gods to use commonplace flint, although those axes will split wood like lightning splinters the sky. No, the leaves of the maguey are as precious and sharp as the tools that chop them down – blades of onyx, greenstone and jadeite, carved with snarling beings, hungry for sustenance. These are the very same tools the Olmec wield to carve their colossal stone heads.

But while Olmec rulers draw power through its spines, most employ the maguey for more domestic pursuits. Ever since their ancestors first moved through this land, the plant has accompanied them every step of the way; even their sandals are woven from its leaves. It certainly helps that the plant produces small vegetative offspring that are easily cut, transported and propagated elsewhere. Over two hundred species cover their valleys. Most of them are wild, although they have domesticated several varieties, which line up around their kitchen gardens to shield the tomatillos, epazote and chayote, defying any animal to pass. Just once every hundred years, so they believe, the plant will burst upwards in a profusion of flowers, a golden inflorescence rising forth to graze the sky. After reaching new heights, it enjoys a fleeting moment of glory before tumbling to the ground and uprooting itself. Yet the maguey lives on through its many children, who faithfully fulfil their mother plant's sacred and domestic roles.

With so many varieties to choose from, one must surely exist for every purpose on Earth. Sweet and nutritious, the heart and stalks are best baked in the oven, served warm and sticky with sap tapped from the flowering plant. Taken internally, the juices treats flatulence, indigestion and dysentery, while a gum from the plant's root is a swift remedy for toothache. Roofs are tiled with the leaves, while the floors of their adobe huts are covered in maguey mats. And after a long day out fishing or hunting, there's nothing like a cup of fermented sap to provide welcome refreshment. Scrape the flesh from the leaves and the fibres are thick and strong, yet silky to the touch, perfect for spinning into thread to make pouches, baskets, fishing nets, hammocks and shoes. While the spindle whorls come out for everyday use, the finest textiles are laboriously woven with slender strands straight from the plant. They find myriad uses for its sharp thorns as well, as tattoo needles, fishing hooks and the tips of spears, although they'll happily leave the penis-piercing to their rulers.

---

The Olmec, whose name means Rubber People, were among the earliest complex societies in Mesoamerica, their influence extending far and wide as they traded rubber across the region. They also blended the substance with different plant juices to make the balls bouncier in their favourite game. After 350 BCE, the population fell into decline, likely due to environmental changes putting pressure on water supplies. But their art, architecture, innovations and material culture lived on, taken up by their neighbours. Indeed, through the brilliant minds of the Maya, the Olmec calendar developed into one of the most precise date systems ever invented.

In 1753, fifty-five calendar rounds after the Olmec ceremonies in their cave, the Swedish naturalist Carl Linnaeus renamed the sacred maguey plant for the Greek word *agavos*, meaning noble, illustrious. While the people of Mexico already had dozens of names for the plant, they could at least agree that the classifier's designation was accurate – more so than some other aspects of his homeland. In the same year maguey received a new name, the tardy Swedish calendar was forced to leap a full eleven days in a single step.

With all our modern technologies and the processing power of computers at our fingertips, it may seem hard to believe that the Mayan calendrical system remains more precise than the Gregorian calendar that keeps all but five of the world's countries in tempo today, and significantly more accurate than the Swedish calendar in the days of Linnaeus. Precision matters, whether in the accuracy of a maguey thorn, the careful tracking of time or the veracity of our stories.

Spinning yarns, weaving narratives – words, like textiles, hold power in their threads. They can restore and raise up, they can shatter and spoil. We must take care how we use them, for accuracy is a responsibility. As I write about Indigenous Peoples and their descendants around the world, I do not in any way attempt to speak for them, only to amplify the traces I hear from their past and stand in solidarity with all who carry their voices into the present. For these are their stories, and they have much to teach us today.

## Appearances can be deceptive

***Denmark • 800 BCE***

In Mexico, they had sharp thorns to endure when harvesting maguey. In Europe, they submitted to an altogether different pain to gather the most precious of plants.

> *Be not nettled my friend, at my praise of this useful weed. In Scotland, I have eaten nettles, I have slept in nettle sheets, and I have dined off a nettle tablecloth. The young and tender nettle is an excellent potherb and the stalks of the old nettle are as good as flax for making cloth. I have heard my mother say that she thought nettle-cloth more durable than any other species of linen.*[7]

Such was the view of early nineteenth-century Scottish poet Thomas Campbell, who lamented the decline of nettle cloth south of the border, a victim of cheap imported cotton. Yet while it fell out of favour in England, the fabric remained popular in Ireland, where the wealthy wrapped their babies in it, and in Scotland, where it went by the name of Scotch Cloth.

The very name of the plant offers a clue to its distant cordage roots. The Old English word *netele*, along with terms in other Germanic languages like *nædlæ* in Danish, bears a striking resemblance to the word needle and traces back to a common root meaning to tie or to bind. The figurative sense of irritation or provocation emerged much later, as reflected in a marvellous saying in the *Dictionary of the Vulgar Tongue* (1785), *He or she has pissed on a nettle*, to describe someone who is peevish or out of temper.[8] From nettle's earliest use for tying and binding, cloth was just a warp and a weft away.

Growing along highways and byways, cultivated and waste ground, grassland and fen, riverbank and mountainside, nettles aren't fussy when it comes to choosing a home, so the discovery of a Bronze Age nettle textile from the Lusehøj grave mound on the island of Fyn was hardly unexpected. Yet analysis of strontium levels – an element in the planet's crust that varies according to geology – suggests the nettles did not come from that island at all; indeed, they may have been harvested as far away as the Alps, transported on long-distance trade routes across Europe.[9] Despite their ubiquity, perhaps few residents of Fyn dared pluck this prickly plant, let alone start experimenting with it. How then did anyone learn that this torturous tingler not only produces fibre, but also yields a green dye? One thing is certain, the technique has been honed over thousands of years.

---

One single-eyed needle, broad and flat, made from antler, wood or bone; that's all they needed to make textiles on the island of

Fyn. Even after they had chopped down the willows, poplars and limes to make dugout fishing boats, paddles or axes, there was never any shortage of bast to make fibre, scraped out with flints between the sapwood and bark. The women would line the fibres up, ends overlapping, and roll them with their hands, sometimes plying them with the tall grasses that grew in the shallows by the lagoon. The fibres served them well, and not just for their clothes: cords crafted from lime tree bast would float upon water, fitting them for any number of maritime endeavours.

That's all in the past. No one is interested in wearing trees anymore; everyone who can afford them wants clothes made from flax. All over the island, they've been digging deep pits, filling them with water to help the plants' wiry stalks shake off their sticky coat. And then they dig more holes in which to build fires, because a maritime climate doesn't make ideal drying weather on the island of Fyn. So, the revelation that nettles are to be used for the great ruler of Voldtofte's funeral shroud comes as a great surprise to everyone. Growing on every barrow, by every pathway, all over the village, many grumble that a wasteland weed is too humble to clad a ruler like theirs, while others can't help wondering who would have the stamina to gather those dragon's spine leaves, their hands seared by the plant's fiery breath, great red rashes flaring up their arms. But no one contests those terrible leaves make cloth smoother than silk, softer even than flax. The women begin steeling themselves for some nettlesome stings, for working day and night to furnish such a shroud – not that there is much difference over these summer months when the sun wheels below the western horizon a few short hours before reappearing in the east.

A collective sigh of relief is heard across the island when it turns out they don't have to pick, spin and weave the nettles themselves. The princely shroud has already arrived, brought to Fyn on an ancient network of trade routes; some even claim to have seen the fabled cloth. The fibres are fine and evenly spun, they report, twisted in the direction in which the sun moves over the earth, while the weave itself is dense and well balanced. It is rumoured to have come from a land that lies far to the south, a place of dark peaks, dewy pastures and lingering mists, providing ideal conditions for retting the leaves. With it came the bronze urn that will hold ashes from the funeral pyre, to be sealed in the pitch darkness of a huge turf mound. Thirty men have been at work for weeks, stripping back grass from the land and piling up turf, the mound growing higher day by day.

Once news of the prince's death spreads, the people of Fyn know it won't be long before those from outside start flocking to the mound. Bringing baskets of food and grain, flints and knives, pottery and textiles, they will conduct their ceremonies and dispatch all manner of grave goods to be buried alongside the remains of their ruler. Amid this great assembly, some will no doubt take advantage by plying their trades, coming by land or sailing in by ship, swapping slaves, repairing weapons and ornaments, trading textiles and dyes, leaving their rubbish all over the place. After paying tribute to the gods and respects to the deceased ruler, everyone will eventually return to where they came from. Then, as usual, it will be up to the villagers to clean up the mess, retrieving what is useful and burying all the smashed pottery beneath another layer of earth.

---

As I delved deeper into this story, I was struck by an unexpected contrast. While the ruler's nettle cloth was soft and finely woven, symbolising a life of prestige and ceremony, the aftermath of such funerals told a different tale. Material debris scattered by mourners and opportunists across the sacred landscape served as a stark reminder of the waste that followed such events. Then, as now, what was deemed unworthy was consigned to the earth, out of sight but never truly gone.

But time has a way of unearthing the overlooked. Amid renewed interest in future materials, the humble nettle may yet find its place. While the road to large-scale commercial production is like bank holiday traffic – a lot of interest but slow progress – a few pioneers are charting the course, including Felde Fibres in Germany, who cultivate nettles using permaculture techniques and follow organic principles.

As the European nettle strives to reclaim its legacy, the Himalayan nettle offers even greater promise. Towering over three metres high, this giant of the family coats the mountain slopes of Nepal, Bhutan and northern India. Although the journey from plant to usable fibre is arduous, it yields a fabric that is both resilient and soft. Not to be overshadowed, ramie, a sting-free relative, holds its place in the nettle lineage. Native to China, where it has been cultivated for millennia, it was named *zhùmá*, the tall grass cloth plant – *má* being the common term for plants that yield fibres. With a silk-like hand that belies its strength, ramie is nearly eight times stronger than cotton when wet and is resistant to bacteria and mould. Perhaps the time has come to reimagine these badland weeds, one sting at a time. Long exiled to the margins, the humble nettle may yet needle its way into the textiles of tomorrow.

## The magic of má

***China • 50 BCE***

In China, the tale of *má* continues, its story intertwined with another plant whose legacy has woven itself even deeper into the fabric of human history. Má was a root word sprouting countless branches. Beyond *zhùmá* for ramie, there was *huángmá* for jute, and *yàmá* for flax. But of all these, it was *dàmá* that stood tallest – *dà* meaning great – a fitting epithet for the generous plant we know as hemp. Over time, people began referring to it simply as má, and everyone understood which plant they meant, for it was no less than a cornerstone of their great civilisation.

---

Legends tell of Shen Nung, the Divine Farmer, born of a love affair between an earthly princess and a dragon of the skies. When he descended to Earth, back in the days when hemp grew only in the wild, he found the world feeble and hungry. With a stomach as transparent as a glacial stream, he journeyed across the land, tasting every herb, fruit and grain, deciphering their secrets as easily as reading the stars. It certainly helped that he could see exactly what was going on inside his immortal form. Then, he began to categorise the plants, ordering them by their roots, stems and leaves, by their medicinal value, taste, soil type and the season in which they would flourish. Of course, he didn't fail to include the marvellous má, which grew in abundance throughout their land. The small brown seeds provided snacks and were pressed into service as oil, as golden as the sun that helped them flourish. Leaves, roots and stems held magical healing powers, curing all manner of conditions, physical and otherwise. And as for its fibres, they were wondrously strong; while male plants provided the finest cloth, the females offered a sturdier fabric more suited to the labours of the day. Shen Nung taught his people how to process the plants and spin their fibres, starting with the basics: hemp string and rope. Once they had mastered these, he showed them how to craft nets to harvest the bounty of the sea, and then he taught them to weave clothing, so they could shield their bodies from whatever the weather threw at them.

Now they can't get enough of the fibre, and hemp's boundless utility extends far beyond the domestic sphere. Nearly two centuries ago, when Emperor Qin Shi Huang commanded the unification of China's defences into one Great Wall, hemp played a crucial role in this monumental endeavour. Not only were his decrees inscribed on paper crafted from mulberry bark and hemp, but the very plaster that bound the stones of the Wall was reinforced with hemp's strength. Even the arrows that soared from the battlements launched from bowstrings of hemp.

The cultivation of hemp has become more than a practice – it is a duty. Although taxes are less onerous than they were in the past, they still have to be paid, and because hemp stems or fabric can be used as currency, it makes sense to optimise their yields. Farmers follow ancient wisdom, turning the soil and planting their seeds with care, following precise instructions set down in early agricultural texts. They fertilise the plants with waste from other life – silkworm excrement, pig manure – returning to the earth what it has supplied. By following such advice, they can triple their yield.

After no more than twenty-five plantings have passed, when the farmers' labours are over and life's final thread cut, their bodies will be wrapped in a hemp shroud, bound with a hemp rope, and laid in the ground. Here, the cycle of life will continue, each generation enriching the soils in which the next crop of má will grow.

---

In hemp, the Chinese people found a reflection of their philosophy, an expression of the interconnectedness of all things. Má wove the very fabric of their existence, the plant's remarkable versatility making it indispensable to a society that valued self-sufficiency and resourcefulness above all else. Over time, hemp would spread far beyond China, travelling the Silk Road, crossing the seas in the sails of ships and becoming a cornerstone of industries around the globe.

Today, as we stand on the brink of uncertainty, hemp reaches forwards, providing a bridge to our future. The construction of tomorrow's cities may well depend on innovations like Hempcrete, a marriage of hemp stalks and lime that upholds the wisdom embedded in the Great Wall of China. Hemp can – quite literally – help to build a more sustainable future. Yet its cultivation is challenging. If nettle is disregarded by many as a badland weed, hemp is widely considered simply to be bad, mostly because of the friends that it keeps. Although distinct from its psychoactive cousin marijuana, both plants are varieties of *Cannabis sativa*, which has led to stigma, confusion and strict regulations surrounding its cultivation. In many countries, the UK included, production is heavily controlled, with farmers required to obtain licenses and adhere to stringent testing to ensure low levels of THC, the psychoactive compound found on the buds and leaves. While working on 'A Textile Garden for Fashion Revolution' at the RHS Chelsea Flower Show in 2022, I was astonished to learn from garden designer Lottie Delamain that we couldn't include hemp in our planting scheme, a reminder of the ongoing challenges facing this ancient plant.

But má has a long history of waiting in the sidelines until its potential is realised. If we dare to embrace it, we may discover hemp isn't so bad after all. Indeed, it may yet reveal the power to shape our world for generations to come.

## Spinning stars, weaving words

*Mali • 20 BCE*

Amid the seriousness and urgency of the challenges we face, we mustn't forget that life, like fashion, can be infused with joy. Sometimes it's through play and creativity that we stumble upon our best answers. As Einstein famously said,

'Imagination is more important than knowledge. Knowledge is limited. Imagination encircles the world.'[10] The Dogon of Mali knew this well. From their home on the banks of the Niger – for at the time of this story they had not yet migrated to their red rock escarpment – the imagination was set free to weave sparks of joy through their textiles, quite literally.

---

Having no cotton to spin that day, the women took down the stars that had been flung into space by the one true god, Amma, and gave them to their children, who stuck spindles through the celestial cinders and spun them like fiery tops. But this is getting ahead of ourselves, because first Amma created two great spirits called Nommo, representing everything twinned and opposite. Flowing and fluid in their green-haired divinity, they could speak from the day of their birth. When the Nommo saw the earth, Amma's first creation, naked and speechless, they pulled ten bunches of fibres from the plants of the heavens so she could be clothed in green, as all women should be. The fibres fell to Earth in coils, as tornadoes, whirlwinds, winding eddies, in an undulating line that carried on to infinity.

Eventually, the day came when the first ancestors were born, although they weren't yet complete beings, each pair embodying the potential for something greater. It wasn't until the seventh ancestor, endowed with the gift of speech and creation, that they were able to fully harness the fibres that had dropped to Earth. At first sunrise, they got up and spat out eighty threads of cotton; it spread between the gaps in their upper teeth like warp threads on a loom. Then they did the same with their lower teeth, so just by opening and shutting their jaws they could make the movement of weaving. In their lower lip was a stud that acted as a shuttle. With the threads crossing and uncrossing, the cloth took shape from their mouth, and the words that they uttered were woven into the threads. The words were the cloth and the cloth was the Word, a continuation of the creative force Amma set into motion at the birth of the universe. In this way, all men and women were able to understand it. It wasn't long before the eighth ancestor was born, and they became speech itself. This is why, when the Dogon sit down to weave, their shuttles gnashing like teeth, grinding out cloth, they call what they make *soy*, which means the spoken word.

Every act of weaving is a reenactment of these primordial events, a sacred practice tying the Dogon back to their divine origins. As they pass their shuttles to and fro, they tell each other stories rich with metaphor about how the earth was born, how she was clothed, and how women and men received language at the same time. They tell these stories to their children too, at least when they aren't running around spinning stars into space. Listen! This is important, they say, because everything that happens in the universe – on the earth and in the sky – results from this movement, the vibrations of primordial threads. This is the very basis of matter, binding heaven, earth and humanity in an unbroken lineage; the life force of all that is – now and to come.

They carry on weaving stories and cloth until the sun dips below the horizon. Then they stretch and get up from their looms. Nighttime is not for weaving: it is for silence and shade. Only between sunrise and sundown can they weave daylight and words into the gaps between threads.

---

As I reflect on the Dogon's practices, I am inspired by the way they intertwine playfulness with profound meaning, reminding us that the yarns we spin, whether literal or metaphorical, shape our world and our understanding of it. The spirit in which they approach their craft offers a valuable lesson in today's frenetic world. Too often our Western lens, clouded by rationalism, obscures our vision of what is always there. Yet in many cultures around the world, there lies a well-trodden path to the otherworld where the boundary between time and space becomes blurred. This is a place where the imagination flies freely, and a creative mind is as important as skills when it comes to crafting solutions – something both Einstein and the Dogon recognised.

I wish I had learnt this lesson sooner. Instead, I learnt it the hard way. For over twenty years, I poured my soul into building my fashion brand, Pachacuti, later channelling that same fervour into Fashion Revolution. The task was monumental, a reflection of the challenges facing the fashion industry; the lines between work and life blurred until they vanished entirely, devouring my evenings, consuming my weekends. Work eroded me like a fault line deep within the bedrock, invisible, undetected or perhaps simply ignored. My foundation began to crumble, until I could no longer bear the weight. I tumbled down, lost in the fall, uncertain of my next steps, knowing only there must be a better way.

It was the morning of my final speaking engagement with Fashion Revolution when the email arrived. Sitting in an Athens hotel room, preparing a talk for The Economist Sustainability Summit, tears flowed down my cheeks: an unexpected vacancy had occurred, not for a new job but for a caravan, nestled in Branscombe's Undercliff. I had reached the top of the waiting list. Without a moment's hesitation, I said yes. I'll sell the house if I have to, I told my husband – and he knew that I would. For this was no ordinary neat-rowed caravan site: my future abode clung to the side of a cliff. Branscombe sits within an irregular landscape of pinnacles and gullies, landslips and undercliff, where peregrine falcons hover high above the inhospitable entrances to smugglers' caves before hurtling down to earth like mouse-hunting missiles. Nestling at the juncture of Triassic mudstone to the West and Jurassic chalk to the East, this red and white meeting of worlds and time is painted across Devon's coastal landscape, and the village lies at its centre. With 150 million years separating the two hillsides, Branscombe's landscape is a visible record of the earth-shifting past.

It is here in this schism between eras that my identity was forged, in the village where my father was born. Now something was tugging me back to the gap between time, to the combe, to the cliffs, to the sea. I knew I could learn from this place, draw on its memories. For it too had a fault line, one that filled with trickling water, abrading, eroding, until one March night in 1790 a ten-acre section of cliff slid away overnight, awakening villagers with mighty convulsions and a thunderous roar. Fields surged towards the sea, taking cows, sheep, rabbits and flagons of contraband brandy hidden from customs men's prying eyes in pits beneath the turf. Hedges and stiles transported through the air came to rest two hundred and fifty precarious feet down, suspended between grassy clifftops and rolling seas beneath. My ancestors must have thought the world was coming to an end that night. Later, they realised God had simply moved their fields a little closer to the village and set about cultivating this newly created valley.

In March, I moved into my caravan, seeking a change I couldn't yet define. It quickly proved both a retreat and a revelation. Around me, the landscape is ever-changing, much like my own sense of self. The fragility of these cliffs mirrors my own vulnerabilities, offering a lesson in resilience. Just as the waves mould this coastline, crumbling its cliffs, reshaping England's beginning or its end, I feel myself evolving, too. In this shape-shifting place, I rejoice in my newfound freedom, in losing track of the days. Time is no longer measured by meeting times but by nature: seconds counted in waves, hours by the tides. I consult the tide table more often than my calendar, a practical necessity for knowing the best time to go shrimping. Sometimes the sea scatters treasures along the shore: sea urchin fossils; black buoys encrusted in mussels that shimmer like a jet choker; and diaphanous Portuguese men o' war, sharp reminders of the dangers of a winter sea. And through it all, Branscombe helps me heal. Like the people of Peru during a time of pachacuti, I have found strength in embracing this state of impermanence, which has allowed me to rebalance and start over again. As the months pass, my imagination soars anew. Here, among the toadflax and viper's bugloss, I rediscover my creativity.

All of us are shaped by the stories we inherit, and the ones we tell ourselves. Yet I have come to realise that we have the power to rewrite those narratives, finding strength through our weaknesses. In doing so, we can restore our bodies, our minds, and even our world. Because sometimes it's in the liminal spaces, where things seem to be falling apart, that the most enduring stories – and solutions – are woven.

# PART TWO
# And Colour Flowed Through Every Land

## Colour as an extension of being

*Papua New Guinea • 30 CE*

What is colour? It's a question I've been asking myself as I write this book. And like most things, there is no simple answer because it all depends on how we see. 'There is no colour in the external world. Colour vision is an illusion created by the interactions of billions of neurons in our brain,' explains an online article in the book *Webvision: The Organization of the Retina and Visual System*.[1] And while this cheerless description may ring true from a scientific perspective, I believe that in seeing colour we do more than observe a spectrum of light: we engage with the palpable currents of the universe that shape our existence.

Colour is the language of light, its spectrum revealing the hidden dialogue between the sun and the earth. Colours are not static adjectives, but a living experience, a meeting point between our solar system and the inner workings of our mind. To see colour is to witness the tango of photons as they collide with matter, reflecting and refracting, absorbed or scattered in a never-ending dance until they reach the human eye. Within our retina, these vibrations are translated into electric signals and transmitted to our brain, where they are woven into the tapestry of our perception. Here, colour comes alive – not just as visual detail, but as an active conversation with the energies that continuously give shape to our world. This is the great revelation of colour.

As much as I wish it were so, plants do not paint themselves for our pleasure, nor to make the world more cheerful. Colour is no accident, nor is it an adornment of nature, but a visual display of the ongoing relationship between life and light, an invisible rhythm that pulses through botanical life. More than this, it is a language of survival that resonates through every vegetal being, each hue recounting a story of connection and purpose – coded messages and vital signals within the entangled web of life. The green that mantles our planet is the alchemist of sunlight, transforming its rays into the energy that fuels life. The yellows, oranges and reds of flavonoids and carotenoids are not just pigments but protectors; responsible for fragrance and flavour, they invite the brush of a bee and guard against battering light. Even elusive indigo is presumed to be a form of cellular protection produced to repel insects or other invaders, revealing its true colour only when it meets with oxygen in the air, much like an elderly relative who, after dipping into the punch bowl at a party, steps onto the dance floor with unexpected flourish. When it comes to indigo, feeling blue is like walking on air.

As we journey through this chapter, we will discover how colours flow across the face of our Earth, charting a path of light, life and survival. Sometimes they bring joy, sometimes sorrow, but every shade, every pattern, tells a story of our shared pathway through the world, mirroring the questions we are grappling with today. And perhaps offering answers. Through this lens, we now travel to the world of the Ő̃mie, a people for whom cloth embodies the essence of being, and colour is not just seen but lived.

---

The Ő̃mie do not call it barkcloth. Stripped from the inner bark of rainforest trees, *nioge* is the only cloth they know, cocooning their bodies, covering their nakedness, absorbing their secretions. The Ő̃mie do not call it barkcloth: they call it their second skin, but it allows much more than sweat to pass through.

Cloth is an extension of being. A woman is a woman because of the patterned cloth around her waist, the material difference that identifies her sex more than anything in her physical form. When a child is born into their community, the first thing they do is swaddle it in well-worn cloth, transferring not just bodily substances, but identity, culture and meaning. Elsewhere on the island, it is the banana they use for cords, ropes and cloth, but not in the realm of the Ő̃mie, even though red banana trees curl elegant fronds over their high green slopes, looking like flame-coloured birds of paradise against the lush foliage. Their use of barkcloth goes back to the beginning, to the place of their creation where the mighty headwaters of the Girua River tumble down the green mountain. This is where the first people emerged.

The woman was called Suja, meaning 'I don't know', a name that hung over her life like a question. In the beginning, there were no children, and no matter how she tried, her body remained as barren as the dry season's earth. One evening, Mina, her husband, came to her with a sharp blade and the suggestion of something she could never have imagined. He made an incision just below her belly and pressed healing leaves into the wound. 'You must go now,' he told her. 'Don't come back until your blood comes. When it does, you will know what to do.' Her days were stretched out in longing, the nights filled with broken sleep from the howls of the barking owl. Then one morning, her blood came. She set off in search of the first tree and set to work, stripping its bark, pounding until it became a sheet of cloth, fibrous and raw, alive with the resinous aroma of fresh wood. But the first cloth was still not complete. She carried it to the river and pressed the fibres into the wet red mud. Now that her cloth carried the mark of the earth's blood, Suja knew she possessed the power to make life. When she returned to Mina, they cut the barkcloth in two: a loincloth for him and a skirt for her, symbols of the life they had yet to make. In time, Suja bore a child, but the barkcloth represented so much more than her fertility: it was the material manifestation of the wisdom of women. A ritual, an act of creation in itself.

All these years later, the making of cloth remains unchanged. To make the mud-brown cloth, the women turn to forest fig, or paper mulberry for those seeking a whiter backdrop on which to initiate their art. The choice, however, is not theirs to make, for only the Duvahe, the chief of the Ő̃mie, can select the design. No one resents this; after all, her appointment is not by wealth, power or birthright, but according to her wisdom in the rites and rituals of the cloth. Standing before one of the many streams that ripple down the wooded slopes, she calls upon the tree to impart its spirit into the infant cloth before making the first cut in its bark, just as Mina once cut into Suja's virgin belly. After this, the rest of the women can get to work stripping the bark and preparing the fibrous material. Once they have rinsed and folded the matted inner bast from the tree, it is time to begin pounding. With wooden club on flat stone, the fibres ease apart, softening and separating with each blow: thud, thud, thud – the sound echoes around the hillside like the beating of war drums.

A strong arm is required, for this is a slow process, taking at least half a day, and it's not until the texture is just right, soft with a glossy sheen, that colour can start to flow.

A woman hovers over the cloth, a song on her lips. In her hand is a betelnut husk, frayed at one end, her chosen instrument for mapping the pathways that will bring lifeforce to the cloth. Black always comes first. (Although, while we may call it black, for the Ő mie this colour conveys a whole spectrum of darkness, even embracing blue as they have no word for that colour.) To make this inky hue, the woman wraps bamboo ash in *tulif* leaves, before chewing the whole mixture up in her mouth and spitting it into an upturned coconut shell by her side. She begins to draw, transferring the elemental forces of nature from plant to cloth. But this cloth, this nioge, is not simply dyed: each line imbues feeling as much as image, layering meaning upon meaning. And although only the Duvahe is able to channel her imagination or visions and birth new designs, this doesn't stop the other women putting their own artistic slant on the process. It is always the same story after all, just told in a slightly different way. Inherited from mothers, grandmothers and back through her clan, these rhythmic patterns are born of many lifetimes of knowledge.

Once the pattern has been outlined and left to dry in the sun, it's time to fill in the spaces between these ancestral lines. Red will come next, a colour that plunges to the depths of the woman's first menstrual blood and soars to the unrivalled resplendence of bird of paradise plumes, not that these creatures are birds at all, but the spirits of their ancestors alive in the trees. She places scrapings of *biredihane* tree on top of ferns from the riverbank, the whole lot rolled over and over as if the fronds have been visited by a leaf-rolling caterpillar. Hot stones from the fire are then placed on top, helped along with a shaking of ash. As the bundle heats up, the woman squeezes it between her hands, releasing the redness of the biredihane. Applied while still warm so it is absorbed by the cloth and doesn't bleed beyond the lines, this liquid is the blood that brings the dark veins to life. Then comes the final colour, yellow from the fruit they call *aré*, picked once it has ripened to just the right golden hue.

These lines are an expression of their clan, of their lineage. Wisdom is what the Ő̄mie women call them. This is colour as an extension of being.

At other times, they will appliqué onto the cloth rather than paint, employing the wing bone of a bat as a needle. When the woman works like this, like Suja, the first woman, she first takes the background cloth to the river and soaks it in mud, squeezing the fibres with her fingers so the clay gets deep down inside. There is nothing dull about mud – as long as it's the right kind of course – from the iron-rich silt by the river. Mud alone won't produce the desired colour, but it fixes and darkens her dyestuff instead, creating a pleasing contrast with the pale layer she will stitch on top. No one knows that this iron is a residue of ancient lava flows. It will be another two millennia before their tree-covered mountain blows its top and the vital life force behind their cloth bears thousands to their death.

Nioge is the Ő̄mie way of seeing, directing their way through this life. As it swaddled at birth, so it will accompany them to the grave, their bodies covered with pandanus leaves and wrapped in this cloth made from trees. This is the very meaning of nioge: the cloth that mediates life. With them from the beginning of time, its images are their books, their poetry, their present, their past. But this doesn't stop them wrapping their sago in it as well.

---

On the steep slopes of Mount Lamington in the Oro Province of Papua New Guinea, nioge has long been more than a textile; it is a canvas of pathways, infilled with colour, reflecting how the Ő̄mie's lives are deeply entwined with the natural world. Born from mineral pigments and the alchemy of plants, each hue carries the memories of the land, the cultural knowledge of wise women and the spirit of the present, stitching Ő̄mie identity to their remote volcanic home. Colour is an expression of their place in the world.

## Things are not always what they seem

***Egypt • 250***

The fashion industry is a powerful force that shapes not only our appearance but our cultural identities. Yet, behind its dazzling designs lies a complex web of practices that often blur the line between authenticity and imitation. Just as the Ő̄mie women of Papua New Guinea create their barkcloth or nioge as an extension of their being, so fashion seeks to weave meaning into contemporary creations. But in contrast to the Ő̄mie's deep-rooted traditions, today's industry often dances with deception, prioritising profit over authenticity and illusion over integrity. The drive to sell more, faster, cheaper, has led to widespread practices of imitation with no threads of connection to the traditions that birthed them.

All this is an echo of a much older tradition of counterfeit craftsmanship, where the value of an object lay not in its material authenticity but in its ability to deceive the eye. True luxury was costly. Perfected in Phoenicia – the purple land – the city of Tyre gave life to the costliest of purples, drawn from the death of tens of thousands of marine molluscs. The desire to emulate drove a buoyant market in impersonations using cheap ingredients, readily available, even if their colours lacked the staying power of the real thing. The ancient Egyptians were masters of such practices, as demonstrated by their recipe books for imitating precious dyes and metals. As we delve into their world, I imagine the Greek scribe who would have been tasked with setting down such recipes, revealing the lengths to which craftsmen would go to replicate the most coveted of colours.

---

A Greek scribe is transcribing recipes onto sheets of papyrus. The priests have assured him that their intention is to enhance the knowledge of Egypt's craftsmen, yet he can't help thinking these documents are something akin to a forger's handbook. Already, he has copied out dozens of methods for colouring gemstones, imitating precious metals, counterfeiting letters of silver and gold, but it's not his place to question, only to write. The scribe is nearing the end of the task now, having just embarked on a series of recipes for dyeing wool. Yet these are no usual colours: almost all of them describe ways to imitate the most problematic of shades.

It is not just in Egypt that purple is a rare and valuable commodity. Across Europe, the colour is elusive, exclusive, divine – all the more so since the authentic way to obtain the famed Tyrian hue is from a foul-smelling gland behind the rectum of a predatory seashell. No one doubts the rumour that ten thousand of the small creatures were disembowelled to trim Marcus Aurelius's robe – holy splendour has its price. Not long ago, Tyrian purple was exchanged for its weight in silver. Now, a pound of the dye is worth twenty times its weight in gold, yet still the shells pile up on the beach to colour the Emperor's new clothes. It was a mania, the scribe thought, this lust for the colour of clotted blood, although perhaps understandable for an emperor, given that Tyrian purple grew richer and stronger instead of diminishing with age.

It had long been decreed that none save the Roman Emperor, his family and senators were entitled to wear the regal colour, but this hasn't stopped men of covert aspirations lusting after what is forbidden. In Egypt, however, purple's exclusivity has softened, tempered by a pragmatic hand. While full robes in the imperial shade remain the privilege of the few, a touch of luxury still finds its way into the ornamentation of tunics and mantles. In dyeing such threads, artisans can stretch the precious dye to its limits, coaxing every last drop of colour from their vats.

According to the recipes the scribe is copying out, one of the choicest plants for imitating malodorous purple is surprisingly common, growing like a weed in the vast deserts to the west. Nonetheless, he makes sure to add a warning against buying alkanet from unscrupulous merchants and *pigmentarii*, for there were certain to be those who would dilute it with pine cones, peaches, beet juice, wine dregs or camel urine given half a chance. The primary issue, as the scribe understands it, lies in the fleeting nature of the purple dye, which will fade quickly unless a suitable fixative can be found. But the recipes he is recording come with their own set of challenges. Take this one, for example. The first ingredient, sheep's urine, isn't exactly easy to collect. Marsh cinquefoil flowers sound relatively straightforward, but in order to dissolve them you must boil pig manure with the urine of an uncorrupted youth – if you can find one – before pouring the mixture over the plant. Despite its strong odour, urine was essential for binding colour to cloth and improving pigmentation. At least the third ingredient is relatively easy, for henbane is already part of the Egyptian medicamenta as a painkiller and sleeping remedy. Once more the scribe feels mightily relieved that he only has to write these recipes down, not test their efficacy.

He starts another recipe for cold dyeing the shade with a warning: *keep this as a secret matter because the purple has an extremely beautiful lustre.* This method involves grinding the scum of woad into alkanet with a pestle and mortar, before adding liquor of kermes, derived from an insect that inhabits the leaves and branches of the oak. As a scribe, the man has little choice in what the holy priests tell him to write, but he doesn't want any trouble with the authorities and everyone knows the consequences of imitating the coveted hue. *You will find the colour beyond all description*, he notes at the end of the recipe, although if by any chance you should wish to make it even more brilliant, you could try cooking the alkanet with purging weed, purging cucumber, wild cucumber or hellebore. As he writes down the names of these pernicious plants, he can't help but wonder at the extreme lengths to which a counterfeiter of colour will go.

Another recipe begins with yet more alkanet, this time with madder roots and orchil from lichen simmered in a pot filled with calves' blood. At times he wonders whether red shouldn't be red and blue be blue instead of all this mixing and manipulation in secretive efforts to outdo the natural world. Some may call it artistry, but he sees it is fuelled by profit and can't help wondering where it will all lead.

At last, the scribe moves on to recreating Tyrian purple, adding the words *Guaranteed Superior Purple* in case any reader doubts its likeness to the real thing. Seven drachmas of alkanet, five of the mineral orpiment, five of quicklime, one of urine and a cup of water is all that's required. While the ingredient list still has its challenges, purple alchemy is, he concludes, all too straightforward when you consider the alternative, and he tries to imagine ten thousand disembowelled sea snails piled up on the shore.

---

In a world where appearances can be everything, this story reminds us that the allure of luxury and exclusivity has always been shadowed by the art of imitation. Today, this issue extends far beyond the counterfeit handbags that swing from our shoulders: contemporary clothing designs borrow heavily from Indigenous and minority cultures. This past summer, I had the privilege of attending XTANT in Mallorca, an exhilarating gathering of textile artisans from across the globe – masters of their craft. Upon my return to England, I took the Tube across London. As I stepped into the carriage, I couldn't help noticing a sea of copies: printed ikat shirts, printed batik dresses, printed embroidery skirts. Often recreated with synthetic dyes and artificial materials, these fast fixes are intended to replicate slow processes. Conjured from lengthy recipes concocted with hundreds of chemicals, they would surely exhaust the hand of our diligent scribe.

This is not a modern phenomenon, as we will see in the following chapters. For centuries, cultures around the world have been systematically looted for corporate gain, their rich traditions reduced to mere commodities. Copyright laws, focused primarily on originality, often overlook cultural expressions, leaving designs rooted in traditional knowledge vulnerable to exploitation. On the long journey towards a more respectful and authentic fashion and textile industry, isn't it time to reimagine how we honour the source of our inspiration and integrate this respect into our design practices? Picture a future where brands go beyond surface-level appreciation, establishing meaningful relationships with artisan communities and engaging in genuine dialogue. Imagine an exchange that is truly reciprocal, helping to shift the power asymmetry that favours the corporate over creators. Credit is given where it's due, fair remuneration is ensured. Collaborations are no longer tokenistic, but genuinely respectful and uplifting of traditional knowledge. In this vision, cultures, communities, artisans and processes become visible in a crucial first step towards rewriting the narrative. Let's look towards that day as we cultivate a fashion industry where imitation is weeded out and authenticity can finally take root.

## The second most noble colour

*France • 300*

In the ancient world, the depth and richness of a hue spoke volumes about a person's social standing. As we have seen, at the summit of this chromatic hierarchy stood Tyrian purple, so close to the divine it was a symbol of the holy on Earth. For those who aspired to grandeur but found themselves occupying a rung beneath the Emperor and his retinue, madder emerged as the second most noble shade. An evergreen climbing plant native to Europe, madder's leaves fan out around the stem like a star. But it is the roots that produce a vibrant and long-lasting red, acting as a passable understudy in the age-old power play between colour and prestige. At times, it even took on the starring role, such as in the creation of Egyptian Purple, a more accessible alternative to the coveted Tyrian hue.

One of the earliest known dyes, madder has been cultivated and prized for over five thousand years, first as a dye and later as a paint pigment. A red thread strung through a bead at Çatal-höyük hints at its ancient lineage, although the origins of that dye have proved hard to confirm. Across the Roman Empire, rivalry simmered among the Gauls, Italians and those of other regions, as each vied to produce the finest version of the dye. The competition centred not only on the richness of the hue – ranging from a deep purplish red to the spilt scarlet blood of a gladiator – but its consistency and quality too. Red was more than a flush of regional pride: it was a cut-throat contest for cultural and economic dominance. Superior dyes fuelled trade, generating wealth, which bolstered the influence and prestige not just of the artisan but of merchants and the elite. We are stepping into a world where mastery of colour held the power to make fortunes and elevate whole sectors of society.

---

The merchant has followed the rich earthy scent all the way to the workshop, although in truth, it isn't until he is close at hand that he can be certain of having arrived at the only sweet-smelling place in the dye district. Odours of rotting cabbage, dung of cows, sheep and goats waft up and down the streets where the *tinctōrēs*, the dye masters, work, not to mention the gallons of putrefied urine they get through every day. Apparently, there is nothing better than this malodorous fixative, and the stronger the better. It is an unavoidable fact that the streets of the dye district stink like a latrine. But pure madder is different. The colour needs just two ingredients: alum, as a mordant to bind the dye to the fabric, and fresh roots of the plant. For Gallic gentlemen such as he – those of discerning taste and smell – the madder dyer is the only place in town.

Not that there are a lot of options, it has to be said. Purple, red and blue are the three noble colours: the darker the hue, the more dyestuffs used, the richer the man. True purple is out of the question for a man of his rank, and, in any case, the exorbitant price of that cloth of eviscerated sea snails is prohibitive, no matter how much land and livestock he owns. If it isn't to be Tyrian, then red is the second-best emblem of luxury. Imbued with sentiments of prestige, ardour, bravery and sacrifice, it is worn by Roman soldiers and gladiators alike, although admittedly their clothes are mostly dyed with *grani coccum*, grains of scarlet, rather than madder. It has not long since been revealed that these grains are not seeds after all, but small insects called kermes that suck on the sap of the oak, not that anyone really cares what they are after having seen the magnificent vermillion dye that they make. But even kermes are ten times the price of other dyes. Madder it would have to be.

Textiles are a visual language even the illiterate – most of the population – can understand, and it is an added advantage for merchants like himself that madder-dyed cloth is less common in the lands to the north. This will surely lend a sheen of exclusivity to his attire. In any case, it makes sense to use madder because Gaul abounds with the stuff, which is no bad thing when everyone wants to wear red; everyone, that is, who is noble or rich. And if the shade turns out a whisper on the purple side, all the better. Not too far though, as the dyeing and wearing of purple is a capital offence in these stratified times. The dyers always know where to draw the line because their heads would be on the block too – and that's if they are lucky. Those who aren't are buried alive.

It is commonly held that the madder of Italy is the most esteemed, particularly from plants growing in the hallowed shadows of Rome. Others argue for the city of Ravenna, a belief which stems from the writings of Greek physician Dioscorides, who praised its madder in his *Materia Medica*. Nearby, the ruddy Rubicon River, named for the iron-rich hue of its silt, may suggest a link to the dye, but as most beyond the esteemed physician understood, muddy waters do not a fine fabric make. What about cloth from Lille? the good people of Gaul argue. Have you ever seen such a hue, vivid as a sunset over the sea? But whoever is right, who can really tell the difference, the merchant wonders, when they put Italian and Gallic cloth side by side? It uses the same recipe wherever it comes from, the one passed down by the Babylonian scribes, for there's little point changing a method when it works so well.

The tinctors first boil the fleece with an equal weight of alum over the fire before preparing the dye vat. Again, an equal weight of madder root to fleece is used – there can be no skimping on either mordant or dyestuff if the hue is to be luminous and long-lasting, penetrating deep into the fibres. Only then will their efforts command a good price. For those willing to risk a shade that veers a little further towards the forbidden, the wool is dipped in a blue bath of woad before following the madder recipe. That's certainly not for him, not when he's heading to Rome. Pure madder it will have to be, as long as the fleece for his new jacket can be dyed in time, for of all the tinctors in the district, it is the madder dyers who do the best trade. Red may not be the cheapest of colours, but there was simply no escaping the truth of the matter: it may come from the soil, but no one treated a man like dirt when he walked out wearing madder.

---

In an age where speed and cost often take precedence over quality, the story of our sartorially minded merchant reminds us of the true artistry and hours of labour that once went into the dyeing of cloth. Madder was more than a colour: it was a living testament to the dye artisan. The merchant's navigation of foul-smelling streets in pursuit of the perfect shade reflects the lengths to which people would go to distinguish themselves through their clothing. Yet while we no longer traipse foetid alleyways in our quest for unblemished colour, those dingy backstreets haven't disappeared, they've simply moved elsewhere. I've seen them outside Dhaka, where effluent flows through open drainage ditches, past food stands where workers buy lunch, straight into the river. Here, red no longer signifies prestige but the peril of the red-listed species of river dolphin threatened by toxic chemical pollution and habitat destruction.

But change is coming. Just as the merchant valued the skills and tradition behind his chosen dye, so we are increasingly seeking out pieces that tell a story of meaningful creation. If this trend were a textile, it would be enjoying a deep soak in the history of artisanal craftsmanship rather than a fleeting brush with passing trends. This analogy extends to how we perceive and value the creative process itself, recognising that true luxury lies not just in the final product, but in the authenticity, passion and craftsmanship that brings it to life. For me, luxury is not a product, not even a lifestyle, but a mindset and a way of being. The word originated from the Latin *luxus*, evoking notions of excess, extravagance and – tellingly – dislocation. By the mid-fourteenth century, it carried connotations of sinful self-indulgence. By the fifteenth century it had evolved to signify sensual pleasure, and by the seventeenth century it had shaken off its pejorative taint entirely, transforming into an aspirational ideal beyond life's necessities. Luxury shifted from an active expression of excess to a more passive, comfortable indulgence.

The merchant's thoughtful selection of madder underscores the importance of skilled labour in creating something truly valuable, a principle that resonates through the ages. When I founded my fashion brand Pachacuti in 1992, I argued that authentic luxury not only delivered a beautiful product but also provided positive benefits to all who were affected by its creation.

In other words, the process of creation is as significant as the product itself. I still hold to that belief today. So, let's reclaim luxury as an active force, and usher in a new era of possibilities for all it can truly be.

## Finding the right words

*Paraguay • 700*

Red was also the colour of the Guaraní, not that they had been given that name yet. In their own words, they were simply Abá, The People. And they were always on the move: making a clearing in the forest, planting their crops, setting up their villages, enjoying their feasts. These meals weren't just about food, although it was surely delicious: they embodied the spirit of community. Everything was shared equally, everyone gave and received, following the example of their first fathers and mothers who hosted and were hosted from the beginning of time. This was the essence of reciprocity. Likewise, when they painted themselves red, it bore no tinge of exclusivity. Red was the colour of the people.

---

The Abá are on the move, having exhausted all that this patch of forest has to offer. They wear nothing but loincloths and mantles dyed red with the seeds of the eternal shrub. In their language, both the plant and the dye it yields are called *uru'ku.* It simply means red – the only red that matters in their world. For the Abá, the uru'ku bushes are living symbols of their origin, and markers that illuminate their path forwards, while the dye they press from its seeds becomes the living embodiment of Ywy Marã Ey, the Land Without Evil, a place unblemished by the scars of time. Their bodies and faces are painted with it too, but it's not just the usual marks they wear when setting up camp. Now they are painted for war, not just against rival tribes or for territorial expansion, but against time itself.

Time is not linear for the Abá: the past is always present, a living memory carried in their stories. But the future – Ywy Marã Ey – is out there somewhere, waiting to be found. To the Abá, this is more than a lost paradise. It is a place of untouched soil on which nothing has been built. A mountain where no one has removed the trees. A place where cotton grows tall and strong without any need to tend it. Somewhere they won't need to kill fierce animals in order to eat, so they can spend all their time feasting and dancing. The only trouble is, they don't know where this land can be found. It might be over the mountains, or even beyond the sea. They only know it exists, and they'll know when they have found it. Because their quest is not only physical; it is a spiritual journey rooted in the very creation of the universe, an ongoing act of becoming rather than a return to what once was.

All this searching goes back to the origin, to the darkness, when their First Father, Ñamandu, tasked Tupã with creating Earth and bringing forth life. From the centre of the heavens to the furthest reaches of the setting sun was Tupã's domain, and it was within this sacred space that Earth was born. This spiritual abode, where the divine and the earthly seamlessly intertwine, is the place they are seeking, somewhere their ancestors have been readying for them since the dawning of time. Their journey is made all the more arduous because creation is always unfolding, just as their creator continues to reveal himself. While the direction remains unclear, they trust the uru'ku and their shaman to guide them towards a future that is always in the making. Unlike most in the great forest, the Abá choose their shaman not for his wisdom or cures, but for his ability to weave words into poetry. Words for the Abá are not static labels but living forces integral to the act of becoming, which is why *ñe' ẽ* means both word and soul in their language. Words sustain life like the marrow in their bones. To speak is to act. The shaman's words are more than a form of communication; they're a continuation of the act of creation, ensuring nothing remains fixed and everything is open to transformation. This means there are none of those barren words some tribes use – a pipe for a pipe, a bird for a bird – but word-souls from the gods, which describe how things look, how they feel. Words that open themselves up into flower.

As they sit around the sacred fire, their faces shimmering with uru'ku, the shaman speaks with the voice of the Mainó, the hummingbird, the first bird to fly through the newborn sky. It is the Mainó who cares for the bird-soul of small children when they emerge from their mothers and open in flower. When the infant's parents go out into the forest, often the child's bird-soul will flutter after them. In a careless morning, a small soul can easily be blown to the skies by a strong gust of wind, but the Mainó stops it getting lost. From time to time he doesn't succeed, in which case he brings word from beyond, from the souls of children carried away for a walk around the sky. As the shaman talks, he puffs thoughtfully on his skeleton of the mist.

Soon the Abá will be moving again, pressing on through the forest in search of the land of their origin, the land of their future. And as they journey, the gleaming red seeds of uru'ku do more than mark their path; they serve as ongoing reminders of life's continuous unfolding, heralding a place where present and future collide in the never-ending act of creation.

---

The Abá's worldview, with its emphasis on becoming rather than being, offers a profound challenge to the frameworks that shape Western thought. This perspective is not an abstract philosophy but is woven into the very fabric of their daily lives, both literally and metaphorically. Their rich pictorial language enlivens the landscape, reflecting a cosmology in which humanity is an indivisible part of nature. Uru'ku is not just a convenient rainforest pigment; it is a reflection of their belief that existence is ever evolving. Every dye, every word, every step speaks to metamorphosis. Just as their textiles are dyed and transformed by the vivid red hue, so too is the Abá's destiny coloured by the constant interplay of change and renewal.

This notion of becoming as opposed to being stands in sharp contrast with the Western perspective, which often favours established norms over transformation. Even our name – human beings – suggests a fixed identity, underscoring our tendency to uphold the status quo. Such a static mindset can reinforce artificial divisions between humans and nature, leading to exploitation rather than stewardship and short-term gains over long-term balance. We see this reflected today in spiralling consumerism that pays lip service to sustainability – and if those lips are wearing lipstick, they might well be tinted with the very same seeds. Uru'ku – commonly known as achiote or annatto – remains one of our most widely used natural colourants for food, cosmetics, paint and textiles alike, signifying our continued dependence on the rainforest and, while the seeds can be sustainably wild-harvested, elsewhere we continue to extract with no regard for renewal.

Everything is in flux, and our survival depends on aligning with the rhythms of our planetary home. As we confront the intertwined challenges of climate change and biodiversity loss, the Abá's worldview invites us to anticipate the future, to see our actions as seeds of tomorrow's world, and to see ourselves as integral to the ongoing evolution of humanity and nature alike. This is the way of the Abá: a path that honours the earth's cycles, recognises that change is inevitable and embraces it as the essence of life.

## There is yellow, and there is great yellow

***Ireland • 1114***

Dyeing and magic are not so far distant, with their cauldrons and herbs, wise women and secret recipes. Colours that appear to be one thing can turn out to be something different entirely, like when earthbound madder and woad hook up to masquerade as the purple of ten thousand sea snails. Such is the case with saffron. Even the word evokes alchemy, coming as it does from the Persian for gold-strung by way of Arabic where it means 'will be yellow' – something that is gold, but is not yet gold all at once.

During the High Middle Ages, yellow carried none of the lily-livered connotations it holds today, and it would be another two centuries before Giotto's Judas Iscariot wore this treacherous hue. Indeed, yellow was considered a passable stand-in for gold, the reflection of divine light. Irish monks illuminated the richly illustrated Book of Kells with a yellow pigment called orpiment. Yellow was eternal, indestructible and it became the colour of Ireland.

Across Europe at this time, the pursuit of the perfect shade was entwined with evolving knowledge, enriching both craft and customs. In Ireland, this fusion of dye and discovery catalysed the art of transforming the everyday into the miraculous.

---

Everyone in Ireland has heard rumours of saffron's magical properties, and not just as a dye. The female flowers of the autumn crocus hold an entire pharmacopoeia within their stigma. The plant is both stimulant and sedative, and while it acts as an aphrodisiac, it also treats the unwanted side effects of desire: physicians prescribe saffron for syphilis, while midwives reach for it to induce an abortion. There is only one problem: saffron has yet to arrive upon Irish shores. Even if it had, few could afford it. Yet this hasn't stopped it becoming the unofficial colour of the land. Being a pragmatic people, the Irish have turned to local plants – agrimony, bog myrtle, buckthorn, gorse, marsh marigold, meadowsweet, pennywort, wood sorrel, broom, dandelion, sundew, water pepper, yellow wort and a dozen besides – to create a whole spectrum of yellows. Tinged with the earth, sap and sea, these dyes are all quite beautiful in their ways, but none equal weld for its intensity of hue. In town and country alike, men and women go about in long-belted *léinte* in all shades of yellow, and all of them saffron, in name at least.

Unlike their dull English neighbours, the Irish love bursts of colour, and the brighter the better! For a linen that bursts with sunshine, the cloth must be bleached before dyeing, a process far from straightforward. Each family has their own secret recipe: fermented bran, buttermilk, lye, wood ash, urine, even sheep and cow dung can be called into use. After boiling the cloth in this odoriferous mixture, they lay it out on the grass and pray for the sun. Since the century turned, those prayers have become more fervent than ever. Of course, the Irish know how to contend with the vagaries of the weather, adapting their recipes to ingredients that are cheap and ready to hand, but recent years have tested even the most stout-hearted among them.

Terrible signs have appeared in the skies, as if the gates of hell have opened, and all the while deep rumblings like bodhrán drums rise from the bowels of the earth, on and on, as if the world itself is being reborn. Even the monks are rattled, warning the end may be nigh. Then just when it seemed people couldn't take any more, last year turned so hot and dry they were able to wade across the deepest of rivers. At last, the flax ripened, calamity was averted and everyone looked forward to a new set of clothes. All over Ireland, fields and riverbanks were covered with cloth. Some years it takes four, five, even six months to bleach, but if this year sees another fine summer, it could whiten in a month or two. Later, as the nights draw in, they'll spread their linen on the frosted grass in the hope it will be whitewashed by moonbeams. No one fully understands how this works, but in Ireland magic is part of everyday life.

Once the linen has whitened, the alchemy of the dyebath begins, the transformation being all the more remarkable because weld doesn't look capable of producing anything spectacular, being an unprepossessing, weedy-looking plant with tall yellow-green spires. But it is often the quiet ones that have hidden riches inside. Add stale urine, oak galls, wood ash or burnt seaweed, and this wasteland weed becomes quite the attention-seeker. It's little wonder they say, *Chomh buí leis an mBuí Mór.*[2] There is yellow. And then there is great yellow.

This is their saffron. This is weld. This is the greatest yellow of all.

---

Plant dyes are as much about transformation as they are about hue. In the early twelfth century, when tensions over political and territorial control escalated and Ireland teetered on the brink of foreign domination by England, the transfiguration of the everyday into the extraordinary became a symbol of audacity. With every bold yellow that emerged from the dyebaths, the Irish were not just clothing themselves in colour; they were emblazing a story of survival in the face of an uncertain future. The weather tested them, the land challenged them, yet even in the darkest of days they found ways to transform their clothes into a halcyon hue.

I relate deeply to this story, and not just because of my Irish ancestry, which is colourful in an entirely different way! When the skies around me seemed threatening, when I found myself recovering in my coastal caravan, I took a walk up the cliff behind me one day and discovered a small patch of weld. No, not a patch, a circle, like a magic fairy ring. There wasn't enough to gather, and in any case I try to hold to the traditional wisdom of never picking the first plants you come across. As the weeks went by, I grew curious about the green spires at the top of another field. My dog, Romero, would hurtle up there on our end-of-day walks in his futile attempt to chase the rabbits who, in the stillness of evening, thought it safe to nibble their green supper blades. I dismissed the plants as ferns, but one blustery morning when the plumes of pampas grass in my garden were fluttering like flags, I decided against my normal walk along the beach and headed up the cliff for a closer look. To my astonishment, they were indeed weld – and there were dozens of plants!

I returned to my caravan, my face bright and wind-kissed, blown out of shape, a few dozen spires nestling in the safety of my basket. The sky was the same grey as distance over the sea. Foamy waves battered the shore – it didn't feel like June at all. But in a bucket on my caravan deck, magic was happening: the sun was coming out. I was dyeing a gorgeous dress, which designer

Anna Mason gifted me to wear on our Textile Garden for Fashion Revolution at the RHS Chelsea Flower Show. The dress was made from a vintage linen tablecloth and was originally dyed by my friend Kate Turnbull of The Secret Dyery using daffodils and dandelions. During the gala evening, a few drops of lemon juice from an oyster dripped onto the dress. I thought nothing of it until a day or two later, when some pale spots appeared; although they weren't overly noticeable, it was a good opportunity to test out my dye skills.

Natural dyes were something I knew about from reading and watching, not doing. Despite my historical knowledge of the art (I was once awarded a fully funded PhD to study natural dyes and the symbolism of colour in the Andes, but I left it behind to pursue Pachacuti) my practical abilities were limited to say the least. Undeterred by the lack of a metal pot for simmering the dye on my caravan stove, I simply placed the chopped weld in a large plastic gardening bucket, poured boiling water over it and left it to steep. Then I sieved the liquid through my shrimping net and added the dress. Some neighbours dropped by, we opened a bottle of wine, and I promptly forgot about my dress until gone midnight. Under the stars, I fearfully pulled the bundle out of the water, expecting calamity, and hung it up to dry. Yet to my astonishment and delight, sunrise revealed an inconceivably rich mustard-gold, not unlike Irish saffron. Dyeing my dress felt like an act of hope.

Drawing such a brilliant yellow from cliffside weld helped me recognise that opportunity can lie in the most unlikely places. In the same way, when the Irish laid their linen on the grass beneath the sun and moon, they weren't just bleaching fabric, they were capturing the essence of the earth and the sky to create something sublime. An augur, a promise, that whatever the heavens or the English threw at them, the fabric of their lives would be gold-spun.

## White is the colour of snow

### *Japan • 1192*

What is colour? When I posed the question at the start of this chapter, I wasn't prepared to encounter a world of such contradiction and depth. If we look at the extremes of the colour spectrum, black is the colour of complexity, of light swallowed whole. With its dense concentration of pigments, black absorbs everything and reveals nothing. White is the polar opposite, light unfiltered by pigment. Some call it the colour of simplicity, while others argue it isn't a colour at all but the absence of absorption, a reflection of light's full spectrum scattered by air. In the end, like so many things in life, it all comes down to the way that you see.

As I think about colour, my thoughts turn to Dyddgu Hamilton, one of the literary relatives who pepper my family's past. Not that I ever met her – she died before I was born – but her multicoloured life, including her pioneering role as one of Harley Street's first female doctors, and later a surgeon, left its imprint on how I see the world. Dyddgu's life was a rich collage of contrasts, much like the spectrum of light itself. She counted such literary giants as Hilaire Belloc, Edith Nesbit and G.K. Chesterton among her friends, and while her work was deathly serious, her own writing – like the mischievous escapades of Cockie, her slanderous cockatoo – could be riotously funny. After she hung up her doctor's bag, she applied her surgeon's skills to rescuing injured birds and tending sick animals as she wandered the woods. People thought her eccentric, perhaps because she saw the world in ways others could not, like her friend G.K. Chesterton, for whom white was indeed a colour. More than that, it was one of great complexity, a shade that embodied the fullness of life rather than its absence. In his marvellous book *Tremendous Trifles*, he wrote, 'White is a colour. It is not a mere absence of colour; it is a shining and affirmative thing, as fierce as red, as definite as black … God paints in many colours; but He never paints so gorgeously, I had almost said so gaudily, as when He paints in white.'[3] Chesterton drew parallels between our perception of white and the way we see virtue, which was not, in his eyes, the mere absence of vice or the avoidance of moral danger, but a vivid and separate thing, like pain or a particular smell. Like healing injured birds. But I digress, and circle back to our question of what white truly is.

What does it mean to live in a world painted white? To see it not as an absence but as a presence, not as simplicity but as fierce complexity? Perhaps there is no better place to explore this than in the snowbound villages of Echigo, where the colour dominates the landscape for eight months a year. It is here, in the remote reaches of Japan, that its true essence reveals itself. Here, snow doesn't simply fall, it transforms everything it touches, forming an undulating quilt that envelops all life beneath. And from this snowfall, the villagers have woven a textile legacy as brilliant as the landscape around them. This is the story of how the women of Echigo turned the unremitting whiteness of their world into something dazzlingly alive.

---

The cloth is a child of the snow; not that it had any choice in the matter. This is Echigo after all, where the cold Siberian wind drops its icy burden. In the mountainous valleys of Uonuma-gun, the villages could vanish beneath a white eiderdown within a day were it not for the wooden barricades shielding their homes from the drifting snow and relentless winds. Every morning, the first task of those who remain through the winter is to carve a frozen staircase to the roof and shovel off the night's accumulation so the house doesn't collapse. But this is an inhospitable place in more ways than the weather; the very name sends shivers through the rest of the nation, for Echigo is where the rascals and radicals hide out. Few in Japan dare venture north of the mountains.

Eight moons: that's how long the snow lasts; and while many of their men head south for winter work, the women are left alone with what nature has provided. And time on their hands. It is the ramie from neighbouring Fukushima Domain that roots them in place; without the dampness of their life below snow, the women would never be able to split the soft nettle stems with their fingernails until they are finer than a strand of hair. In the dry southern winters of Edo, these threads would become brittle and break on the loom. This cloth is born of solitude and snow.

They've been making it for so long that no one can quite remember who designed so elaborate a process, but with fifty, sixty, even seventy different steps, there's little chance of boredom in these long-drawn-out winters. The cloth absorbs all their creative and spiritual energy. Especially now Minamoto Yoritomo, head of the Shogunate, has commissioned one thousand bolts of Echigo cloth to present to the imperial court, all of which needs to be ready by the seventh month. Not long ago, this cloth was made only for trading and tributes, but now it has become fashionable; in the sultry summers of Edo, the puckered texture of an Echigo kimono feels fresh as the first breeze of autumn.

Each weaver is responsible for three or four *tan* – one tan being the length of a kimono or the height of eight women. But everyone in the household plays their part, and even the men who stay through the winter will help with the housework. The youngest girls aren't entrusted with weaving these fine fibres yet. For now, it's enough to watch and learn. Only once a girl has come of age at thirteen can she weave her first piece of cloth, and not before she has completed her Jusan-mairi pilgrimage to the weaving deity to pray for knowledge, happiness and health. She will learn quickly after that, for although Echigo is famed for its beautiful women, everyone knows that weaving skills are more important than looks when an eligible young man is on the hunt for a wife.

Back in the tenth month when snow first started to fall, the ramie fibres were shredded with fingernails, moistened with mouths, manipulated with hands and the ends of each strand twisted into the next to form a long and continuous thread. Now, it is time for weaving. Sitting on a plank of wood, legs outstretched, the wooden loom low to the ground where the humidity is highest, the weaver shifts her weight from side to side to get the tension just so. The secret to good work is to feel close to the threads; each has been rolled by hand, so each has its own character. Weaving Echigo cloth cannot be rushed. Two or three moons will pass before the warp threads on one complete tan can be snipped. At the end of the cloth, they cut a narrow strip and offer it at the shrine along with their prayers – a slender banner of thanksgiving for their health, safety and fine weaving skills. At least two more tan will be needed before the snow melts.

When the weaving is finished, the cloth is still far from complete. Water must be collected from one of the meltwater rivers cascading from the mountains; once warmed, they plunge in the cloth and trample it beneath their feet. It's not worth skimping on this stage, because a good massage is what makes the ramie lustrous and soft. Finally, when the weather starts to warm, they lay long bolts of cloth side by side over white-draped fields, the bleaching power of spring sunlight intensified by the evaporating snow. Two or three weeks later, the beige stripes have turned white. And that's how the women of Echigo make snow cloth.

Of course, some will argue that white isn't a colour at all, to which the villagers respond that it is the colour of their world for eight moons a year. And after all, this is not any pedestrian shade: this cloth is the vivid white of an Echigo winter. They've made it that way for hundreds of years and will continue for hundreds more. As long as there is snow.

---

For the villagers of Echigo, white was a luminous, affirmative colour in its own right, as definite as the snows that shape their world. As I thought about the transformative qualities of snow, it reminded me of a story Orsola de Castro, my fellow founder at Fashion Revolution, told me as we walked from the train station to her house, a place where passionate plans took shape, helped along by steaming bowls of pasta. One winter day, Orsola got lost in the snow. This was no strange landscape; it was the same route we were walking. Yet the short distance had grown vast and unfamiliar. Like an Echigo snowfall, well-trodden London streets lay beneath an eiderdown of white. Disorientated, she called her husband, who guided her home. For a time, her family worried: how could she have forgotten the path home? It turned out that Orsola, like the Guaraní, navigates by colour: left at the red bush, right at the green tree. When the world turned white, she quite simply lost her way, just as the Guaraní today are losing their homing signals, the red *uru'ku* berries, amid runaway deforestation.

While we may not rely on colour in the same physical sense as Orsola and the Guaraní, symbols rooted in our surroundings still guide us. When the world around us changes shape, whether through snow, deforestation, floods or any other aspect of our changing climate, those visual cues can vanish, leaving us lost in a landscape we no longer recognise. And I wonder, could our plant-coloured past offer more than beauty? Perhaps it too is a map showing us the way home – not to a physical place, but to a new way of being.

## Colour wars

*England • 1237*

While whiteness reigned in the realm of Echigo, thirteenth-century England was a land of contrasts: the tranquil greens of the countryside belied simmering tensions roiling beneath. The medieval landscape was more than a patchwork of wild spaces and cultivated strips; it was a canvas where the fortunes of nations and the lives of common folk were intertwined. And it was a seedbed for a deeper, more elemental force, one that drove the very essence of life itself – a force just beginning to find its voice. As it stirred, a new word took root in the minds and mouths of the people.

---

A word that means birth has just been born. Derived from the Old French, meaning the principle of life, the word 'nature' grows quickly, gaining meaning and power. Before long, it is being used for the restorative energies of the body and the capacity for growth. Soon it will sprout into the forces that produce all living things, the essence that drives the material world. From there it will spread into the wild. The dictionaries of Europe can barely keep pace with the expansion of nature.

Within nature's embrace, everything soon finds its place. The human, of course, stands at the pinnacle of the great chain of being, while near the bottom rests the vegetative world, only trumped in its lowliness by mushrooms, mosses and unmoving minerals. Plants, lacking agency and voice, serve no purpose other than to benefit sentient beings. Nonetheless, even within their lowly category, some plants are naturally more esteemed than others. When it comes to those used for dyes, purple and red have long dominated the colour hierarchy while blue has largely faded from Europe's palette since Rome's conquest of England. For the Romans, the colour symbolised either the sky blue of the gods or the sea blue of barbarians, and everyone knew into which stratum woad fell. But the ocean is an unpredictable place; sometimes it spews forth riches that are long thought lost. Such is the case with blue, which rises again from the savage sea.

France is the first to reclaim blue as its royal colour, paraded across crest and clothing by King Louis IX. In no time at all, the spectrum is reworked. No longer does the colour embody the uncultured, the uncivilised; how can it when its latest incarnation is as the favoured shade of the Virgin? Violet and grey, brown and dark green are banished from her wardrobe and even her blues start to brighten in French artists' hands. Blue is purity, blue is righteousness and blue is fashionable for the first time in over a thousand years. It isn't long before the rest of Christendom starts following Saint Louis' sartorial lead.

When Henry III starts wearing blue too, woad's bright yellow blooms start spreading across the English countryside. But although woad sprouts up easily, transforming the plant into solid blocks of dye is a long and skilled process. The leaves must be picked – five cartloads an acre if the farmer is lucky – then pulped, drained, hand-rolled into balls, dried, pulverised and fermented into a steaming, stinking mass. This process alone takes three months or more. Finally, the woad is dried again and packed into barrels, where it can be aged for up to four years to increase the intensity of its hue. Dyers are struggling to meet rising demand, so they start importing the stuff from abroad.

The City of London grows rich on the import duties, but the merchants of Picardy are none too happy. For the privilege of landing the best woad in the land, they pay hundreds of pounds into the city's coffers – only to sell it from wooden slats of the quay. Eventually, an agreement is reached allowing them to store the dye in one of the warehouses lining the north bank of the Thames and distribute it around the country. In return, they pay fifty marks sterling a year during the great fairs of St Ives, Winchester and Boston, not forgetting their generous contribution towards an underground channel to bring clean water into London. Not to be outdone, the merchants of Amiens soon make a similar agreement with the woad-smitten wool town of Norwich. Little by little, England turns blue – or at least those who can afford it do.

It, of course, comes as little surprise that madder merchants are not best pleased with woad's rise. All across Europe, violent clashes break out between the red and the blue. In a desperate attempt not to be outwitted by the Virgin, the merchants of Germany rally sacred art to their cause. In frescoes and church windows, blue swiftly becomes the colour of hell, the devil, pain and damnation.

But fashion wins out over religion in the end, and barbaric blue continues its ascent to the heavens.

To support its rise to the upper echelon of plants, woad has healing properties, too. It is the first remedy reached for in cases of pox in the eyes and, if a young child should happen to touch their hand to the fire, there's nothing better than woad powder boiled in butter, not that madder isn't said to be equally effective in healing wounds and treating other disorders.

Of all woad's hues, the most desirable is the one the cloth merchants call perse, fetching six shillings a yard. Perse starts off as a bluish grey, but as no English dyer has ever visited the land of the Persians from whence its name derives, the colour changes hue like a winter's sky, growing ever darker until it settles on a purplish black. Perse must be dyed on pure white wool using woad and ashes alone. But woad is not easy to work with; it is not even a dye in the traditional sense. As a dried pigment, woad must ferment in a dyebath at a steady temperature with just the right quantity of potash for two to three days. Only then will the dyer achieve the correct shade of perse. And the hue will only be true once the cloth or fleece is exposed to the air; while soaking in the dyebath, it remains steadfastly green. Inexperienced or hasty dyers might try to correct their deficiencies by plunging woad-dyed wool into a madder bath; some even dispense with the costly pigment altogether and use black or grey wool with madder and alum in a practice strongly condemned by the guilds.

Beyond their conflicting hues, there is another marked difference between madder and woad: their impact on the land. Madder's deep roots loosen the soil, enhancing the land's ability to absorb water. At the same time, it disrupts cycles of pests and disease, allowing other crops to flourish in its wake.

Woad, meanwhile, is a demanding crop that exhausts the land, stripping the soil of nutrients, leaving it incapable of supporting even the most basic vegetation. Where woad has been sown, even grass will not grow. As woad spreads across the countryside, it leaves a trail of desolation, exacerbating the struggles of those who can least afford it. If only woad and madder could be grown in alternating cycles with other crops, perhaps the burdens on the poor would be eased: the land could be both productive and profitable. But no, there are madder growers, clinging to tradition, and woad growers, driven by the lucrative dye market. Their interests are as divergent as the colours they produce, and no amount of earthly reasoning will blend these opposing hues.

---

The tale of woad and madder is more than a story of medieval England – it could be a parable for our times. Just as the poor reaped a bitter harvest from woad's insatiable demands, so the marginalised today bear the brunt of decisions made far from their lands and homes, wherever they live around the world. Our modern landscapes are still battlegrounds where the interests of the powerful clash with the needs of the vulnerable, and where lowly plants are dismissed and uprooted without a second thought. Our modern political landscape is also polarised, with opposing factions of red and blue clashing over the direction of our shared future.

Yet, within this story exists the recognition that balance is possible, that our earthly home can be both nourished and productive. Natural fibres and dyes are ingredients in the future recipe of our clothes, but they are not the whole answer. Yes, land can grow fibre and colour, but it must also grow food. The key is not to swing from one extreme to the other, but to consider the full spectrum of possibility. Because in the end, the colour wars between woad and madder are not just a relic of history but a reminder that the choices we make today will leave their mark on the world just as indelibly as the colours on a dyer's cloth. If we can move beyond rigid binaries, beyond the stark contrast of primary colours, and embrace the art of blending, of nuance, like the rotation of crops that enriches the soil, we may yet design a future where prosperity means flourishing for all.

## The importance of balance

***USA • 1280***

As the word nature was taking shape in European minds, the Aniyvwiya had long held their own understanding of this primal force. In their misty, mountain home straddling present-day North Carolina and Tennessee, nature was no abstract idea but a tangible presence, as real as the black walnut trees towering at the forest's edge. Earth, water and sky, along with all creatures and plants that inhabited these spheres, were not seen as others to be mastered or conquered, but members of their community who must be sustained. After all, it was through their collective work that the Aniyvwiya came to have land at all, from the turtle who brought mud from the seabed and the buzzard who dried it into mountains with a flapping of wings. While Europe grappled with the expanding notion of nature, the Aniyvwiya simply lived in harmony with it.

---

In the Land of Blue Smoke, black walnut trees are some of the tallest around, standing dark and tall at the margins of the forest among oaks, hickory, black locust, eastern hemlock and sweet birch. Higher in the mountains where squirrels have scattered its nuts here and there, they blend with white walnut, butternut and the eternal blue mist. This is where time began. This is where the real people live. Aniyvwiya is the name they give themselves, although others of a foreign tongue will call them Cherokee, the people of different speech.

For the Aniyvwiya, every part of nature is alive, which means every part of nature has a voice. No voice louder than any other. Their creator put everything on Earth at the same time, so all are equally valued. This is why, when the women go out to gather plants for fibres and dyes, they take only what they need, not what they want. Nor do they take what is easiest, passing by the first plants and collecting only from the second or third, so the first source remains for generations to come. Yes, it may take them a little more time, they may need to walk a little further up the hill, but they have an obligation to keep the system in balance if they are to continue enjoying the good life. This is also why, when they reach for the wolf-bone handle of their knife, sharp as the winter wind, men and women give thanks for the life of the plant, chanting and singing in gratitude. In doing so, they connect with the plant, for when you speak in Aniyvwiya you see and feel what you're talking about.

The women are selective about plants in other ways too; they have to be when they're dyeing, for every shade tells a story. Red is the colour of the wind from the east, of blood, triumph, power and good fortune. It's their favourite hue for clothing and for dyeing their woven baskets of river cane and white oak. Once in a while, they'll decide on a pinky red after gathering inky berries from a patch of pokeweed, but their usual shade is from bloodroot, dripping red like the open veins of the earth. A small piece is sufficient, lifted from the leafy cushion of the forest floor before the plant dies back in the autumn, and no one needs to be reminded of the importance of leaving enough root for regeneration to take place come the spring.

White holds a special place in their minds; the colour of a southern breeze, it breathes of the fullness of the natural world. Tassels from the white-tailed deer, plucked from under their tail, mantles from white turkey feathers, beads from the white river clay – nature needs no modification where white is concerned. Not like its opposing force, the north wind. Blowing itself blue, it brings trouble and defeat. Blue is a hue best avoided.

The west wind is black, the wind that bears death. The death they fear most is the world wearing out when the cords that suspend it have worn thin. One day, those cords will break, the earth will sink into the ocean, and all will become water once again. But until that time comes, it is sickness, old age or war that will claim them, and for that they can prepare. Whenever men ready themselves for war, they prepare a black pigment to paint the story of their life across their bodies. The roots of the black walnut produce the deepest dyes, like the coat of the panther that prowls their camp by night, but the weavers know that stripping the deep roots kills the tree. The men only cut down black walnuts and dig up their roots when they are growing close to their settlements, for no crops can thrive in their shadow. Like a possessive lover, the tree releases juglone, a toxin that claims the soil for itself, refusing to share space. Along with black from the roots, walnut's deep-furrowed bark and yellow-green leaves will dye browns and greens too, but the richest hues are from the shiny green orbs that fall from the trees when summer swings into autumn. Most will be gathered; some will be left to rot in the tannin-black soil that circles the base of the tree.

The nut's bitter tannins are a blessing for the women who now have one less task to worry about, as its presence eliminates the need for a mordant, helping the dye cling to the cloth. Freed from this chore, they can focus on their other responsibilities: carrying water, coiling pottery, weaving baskets, making clothes, tending gardens or caring for the sick. Ochre lies in piles, waiting to be pounded into fine powder for painting the walls of their caves, inspired by the spirit creatures whose presence lingers in scratches on walls, soft treads on the floor. There are children to raise, too – those of the whole community, not just their own. And always the corn, its bright yellow kernels bursting in rhythm as they are pounded down to powder and leached with lye. Soon the air will be pungent with the sweet smell of hominy, blending with the breath of the forest, the vapour rising from every tree that shrouds these mountains in a perpetual blue haze. This is why they call the place Shaconage, the Land of Blue Smoke.

---

For the Aniyvwiya, the black walnut tree was not simply a source of material wealth, but a symbol of how imbalance could sap life from the land. Where every element was interconnected, every action had its consequence. As they managed the balance between the walnut's dominance and the needs of their crops, they knew that survival sometimes demanded the careful removal of what stifled growth to make room for new possibilities. In the same way, this story gently asks what are we prepared to part with in order to nurture a future where all life can flourish?

While Indigenous Peoples have uniquely rich and ancient relationships with their lands, which must be respected and protected,

I believe all of us can forge a deeper bond with our patch of earth. I am privileged to feel a profound connection to Branscombe, where my father's family has lived since records began, but I don't believe the way I feel about the place relies on generations of rootedness. A relationship with the land is something all of us can consciously cultivate wherever we are and regardless of how briefly or how often we call a place home.

## The birthplace of the rainbow

***Peru • 1290***

In the heart of the Inca Empire, where Cusco was considered the navel of the world, the centre of their known universe, colour was more than a visual experience; it was a sacred language. This was no ordinary dialogue, but a conversation with the cosmos, a celestial thread binding textiles to ancestors, nature and the divine. Every shade was a word in this numinous lexicon, carrying meaning that stretched far beyond the material world.

While Inca rulers in Cusco claimed the rainbow as a divine emblem of their authority, the women of Chinchero were conjuring rainbows of their own not from the sky, but from the earth beneath their feet. Coaxing colour from roots and leaves, bark and blossoms, every hue held a purpose, a connection that bridged the seen and unseen. And with these chromatic threads, they wove the ethereal into the tangible, creating cloth that recounted the stories of their world.

---

A silver mist rises from the courtyards, melding with low grey clouds, heavy with rain, but everyone knows it won't last long.

Chinchero is, after all, the birthplace of the rainbow. Look east towards the snow-capped peak of Sallqantay and you'll see one arching through the sky from a field of potatoes. Turn west to Huaypo lagoon and there's another springing up through the maize. Of course, there is a strong temptation to pause and marvel at their fleeting formation, yet when the magic of the sky weakens and coalesces into colour, this spectacle can prove perilous; young women even avoid glimpsing their reflections in the glassy lake for fear of being impregnated by the rainbow spirit. But it's not just in the sky that rainbows are born. On the high Andean plateau, the thin air is suffused with the gentle clacking of backstrap looms and the steam of a dozen dye vats.

Red comes from the fermented roots of *chapi*, a small leafy plant that creeps over the ground like a vine that's got lost on its way to the sky. Other times, the women use green leaves of *thiri*, or the sap of *chica*. If a duller hue is called for, perhaps the deep red fruit of *mote mote* will do, dried and ground into a powder. To darken the colour, the stagnant iron-rich muds they call *mio* and *patchu* are the best place to go; sometimes they leave their textiles buried for days in these layers of ancient sediment, soaking darkness into their threads.

For orange, the women gather the bark of the slow-growing *yanali* tree, shredding it into pieces and dropping them fresh into the steaming pot. If they're looking for a more russet tone, they might chop up *nanci* bark, letting it steep for several days.

Yellow is the colour of maize, of gold, of the sun, one of the most important hues in their rainbow, yet the women can't help but wonder whether their tendency to weave all shades of yellow stems just as much from their fondness for virtuous *molle*. This tree grows abundantly in the highlands and every part of it yields dye. But it does more than offer brilliant colour. Its white resin embalms the dead, the rotting leaves make a fertiliser that improves the corn harvest and its ashes are surprisingly effective as a mouthwash. As if this isn't enough, the peppery seeds flavour everything from beer to honey, curing sore throats, headaches, toothache, constipation and rheumatism too. On top of these myriad benefits, when it comes to dyeing, there's no need for another fixative: simply tip molle ashes into the dye vat along with the leaves and bark, and the cloth will be as bright in a generation as the day it was born. If pushed, the women might try out other yellows from time to time. Perhaps the flowers from the small *q'olle* tree, which turns yarn a zesty lemon or deep mustard depending on how long they leave it in the dyebath. If they've missed the season for q'olle, then the yellow from *ñuñunqa* leaves is equally bright. Or maybe they'll turn to *q'aqa sunka*. There are just too many choices for yellow, but they usually return to molle in the end.

All shades of green come from *ch'illca*, a straggly bush that grows in profusion along the riverbanks, which is a good thing because they need a lot of leaves for a rich deep green, the colour of their fertile plain. Nothing is wasted: its branches make baskets and the ashes help activate the coca leaves, which they chew as they work. They make sure to leave enough of the plant standing, because ch'illca is important for the health of the soil.

Blue comes from *tara*, a bean-like pod from a small thorny shrub. If they boil these pods long enough with putrefied urine, it is one of the easiest ways to get close to black. There's *kinsa k'uchu* too, named for its three-cornered leaf. *Huk*, *iskay*, *kinsa*… The women point to each corner when teaching their children to count, although it's not the plant itself that yields all tones of blue, but a black fungus that calls it home. As far as the dyers are concerned, these leaves are more than a convenient dwelling for the fungus, as they help fix the dye to the yarn with no need for a mordant.

When it comes to indigo, well, there is only one plant they would consider, continuing a tradition passed down over thousands of years. The small, pink-flowered shrub grows along the coast, near where the dark pyramid once lay, as well in the great forest where their mountains slide away to the east. There's no shortage of goods to trade for this pigment, not when the fields surrounding your village fill the grain stores of the empire. In any case, it's always good to have some indigo to hand as an antidote to scorpion bites.

At last, they come to violet: the first colour of the rainbow, the colour of Mama Uqllu, who gave birth to their first great ruler before settling down to teach women to weave. Violet comes from the cobs and kernels of *kculli*, their favourite purple corn; at least, it does when there's any to spare after simmering it with quinces, pineapple rind and spices to make a delectable drink. There's always the low-creeping *ayrampu* cactus, if needed, and although no one likes going near those lethally long spines, they have to admit the lucent dye from its seeds is second-to-none, not just for dyeing cloth, but to colour drinks and puddings as well. After a long day stirring the dye pot or sitting down at the loom, there's nothing quite like a warm violet pudding and a refreshing beaker of purple corn chicha.

This is how they turn colour into cloth in the place they call the birthplace of the rainbow.

---

As colour exists within a spectrum, like rainbows contained within their arc, the people of Chinchero understood the importance of living within limits, of a beginning and an end within which creativity could flourish. They knew that the abundance surrounding them was enmeshed in a network of mutual gift-giving, not to be taken for granted or exhausted. Each plant was harvested with care, with permission, with an offering, to maintain the balance that allowed all life to thrive, and ensuring the land could provide for generations to come. A world without limits is a dangerous, unbridled place, where the absence of boundaries leads to chaos rather than creation. Yet within the arc of a rainbow, within the natural order that honours beginnings and endings, life can be truly extraordinary.

# PART THREE
# Weaving Shadows and Light

## The world changes shape
### *Bahamas and Cuba • 1492*

Sometimes the world changes shape when we least expect it. We embark on journeys with a destination in mind, but when we encounter the unexpected it can take a while to shift our perceptions. For all his flaws, Christopher Columbus was a master of dead reckoning, a navigational method favoured by Genoese pilots, which relied on the ship's last-known position, along with speed, distance covered and the ship's compass in order to plot a course. This method worked well enough across short Mediterranean distances, but Columbus had set his sights on unfamiliar horizons. There was perhaps a reason Iberian navigators mockingly called the position obtained from such calculations the point of fantasy.

When Columbus eventually made landfall, he attempted to determine his latitude. His calculations were wildly off – and he knew it – points of fantasy proving hazardous over distance and time. Yet he recorded them in the ship's log, nonetheless. What Columbus found was not what he was looking for, but his illusions would reshape the world. His voyage didn't just redraw maps; it accelerated the entanglement of plant materials, colours and cultures in ways that had long been evolving but would soon start moving at unprecedented speeds, carried on the tides of trade, empire and opportunism, setting in motion profound cultural and environmental changes that reverberate through time to the present day. As fibres and dyes became intertwined, their histories could no longer be told apart. This chapter steps into that convergence.

Learning from our past is essential; it is this book's unwavering point of departure. Yet, while this chapter calls for reckoning, it is also a cry for growth and renewal, for we cannot remain rooted in what has been. Just as textiles once reshaped global trade, so too can we write new narratives to reshape our future.

---

A man is readying three ships in the harbour, all resplendent in indigo hemp sails. Like a bird setting off on its first migration, his body twitches with restlessness. Christopher Columbus is eager to begin his quest for a westward route to where the dye for his sails originates: India. At last, he pushes out into the unknown, a map of the stars imprinted on his mind, guiding him towards his destiny. The world he imagines slips away.

In his estimation, the voyage should take no more than a week, for the *Second Book of Esdras*, a scripture found in the Latin Vulgate, clearly states that God created the world in seven parts, six of them dry land and the other part water.[1] Not that he hasn't done his own research to back it up, spending months poring over documents and charts by Ptolemy, Marinus of Tyre, Al-Farghani, Toscanelli, to measure the span of the earth and the distance to India. With each new calculation, the earth shrinks a few more degrees, so that by the time he has finished, the *Mare Tenebrosum*, the fabled Sea of Darkness west of Portugal, has become far less forbidding. In fact, it can't be much further than sailing home to Genoa, and it's sure to be faster with the wind at his back.

A month goes by. The islands so liberally dotted across the shadowed reaches of the Atlantic by early mapmakers turn out to be as fictitious as its gallery of one-eyed monsters. But Columbus doesn't lose heart; if anything, the infinite fires his imagination still further, for he believes he has been chosen to carry out a holy mission, the fulfilment of the prophet Isaiah's promise that *the abundance of the sea shall be converted unto thee.*[2] Trusting in divine providence, he believes untold fortunes will emerge from the untamed ocean. Added to which, he has a reputation to uphold, for hasn't he been prophesying about this westward route for nigh on twenty years? The fabulous riches of Asia must be at hand. As a second month at sea draws to a close, the sailors grow increasingly restless. Maybe they'll slip off the edge of the world after all, or perhaps there won't be enough wind to power the ships back uphill?

Suddenly one night, the sterncastle lookout raises a cry: *Land ho!* A light flickering in the darkness, rising and falling like a guttering candle flame, almost extinguished by the blackness of night. They furl sails and drop anchor. A light breeze through the skeletal rigging sets the blocks jangling, and it carries the verdant scent of the shore. Christopher Columbus' nostrils twitch in anticipation. By the time dawn comes around, he is unfurling the royal banner, claiming this island of the Indies for Spain's King and Queen. With a few thuds of a mallet, Guanahaní becomes San Salvador, two hemispheres collide, and the world changes shape.

Later that day, as the *Niña*, *Pinta* and *Santa María* bob at anchor, they are approached by a canoe, longer than one of their caravels, yet fashioned from a single log. Forty paddles swish in unison through the waves. On deck, the sailors step back behind the strong ropes holding up the main mast. Shrouds, the men call the cords that rise to the sky, the same word as the cloth for covering their dead, although these twists of coarse hemp offer little in the way of protection; they have gunpowder for that. It soon becomes clear that the locals come not with poisoned arrows, but with all manner of goods, eager to trade. Their canoes brim with balls of spun cotton, flame-winged parrots and javelins with a fish tooth strapped to one end. *And they traded them to us for other things which we gave them, such as small glass beads and bells*, records the Admiral in his letter to the Crown.[3]

As the three caravels set off to explore the east of the island, more men and women swim out to their boats, for they are accustomed to trading with visitors who arrive by sea, from near and afar, as long as the exchange appears equitable. See the men who came from over the horizon, they point at the motley crew who sailed in from the hoop of the world. But their words are lost in translation. *See the men who come from heaven*, Columbus writes in his log, ascribing to himself a higher blue realm.

Like a compass reorienting itself, Columbus presses on. There is no chart for navigating this strange land, but doubtless they are nearing the kingdom of the Great Khan. The presence of cotton surely indicates fine cloth is close at hand? From island to island he journeys, questioning without comprehending, observing but ignoring the signs, exchanging tawdry trinkets for cotton, pearls, even gold. Eventually, the ships reach the north coast of Cuba, a place where boats come and go all the time, for an interwoven network of material exchange has linked this coastline for millennia. As soon as he steps ashore,

Columbus finds the white gold he has been searching for: huge quantities of cotton being gathered and spun, all growing naturally in the wild. A single house contains over twenty-five pounds of the stuff. His eyes light up as he does the mental calculations: hundreds of thousands of pounds of cotton must grow here every year.

And that's not all. Up in the mountains, where the forest thrums with life, cotton explodes across the canopy like the puff of ten thousand muskets. The cotton trees bear fruit the size of a nutmeg; they burst open when ripe and are blown by the wind, falling in clouds from soaring trees where the kingbirds nest. But the colonists soon discover the celestial down is of no use at all – its fibres are too short to spin. Undeterred, they turn their attention to the timber, noting how the locals use these trees for their lengthy canoes. As plans for a settlement take shape, cotton-tree wood will serve them well for construction in the years to come. In any case, Columbus knows his salvation is to be found in riches beyond cotton, so on he sails, his compass directed towards mountains of gold. The islanders watch the three masts dip below the horizon like the sun. Bemused, they turn over the hard shiny novelties in their hands, so different from the pottery and stone tools that are their preferred goods of exchange. Perhaps the visitors will learn the language of reciprocity should they ever return. Then they head back to their homes to spin some more cotton.

Columbus returns the following year, bringing men, pigs and more tawdry trinkets to settle an island he named Hispaniola. In the shroud-like darkness, no one notices the passenger that jumps ship; riding in the lungs of swine, it is invisible, silent, save for a cough. But it won't be long before its cries are heard. Westwards, northwards, east and south the influenza blows, like a *hurakan* tearing through their palms. Nobody knows where it will go, but they know where it's been by the long trail of bones.

---

My own sailing journey began not by design, but by chance, an unexpected detour that shifted the course of my life. Becoming a sailor was never part of the plan, but sometimes adventures pull us in when we least expect them. In search of cheap student accommodation, I'd driven my Morris Minor down to Southampton's marina to enquire about renting a cabin on a decommissioned sewage barge. The vessel was empty of life, so rather than waste a trip, I wandered the waterfront. An hour later, I found myself learning the ropes (which I soon discovered were called sheets and lines) on *Velsheda*, a sleek 1933 J-class with the tallest single mast in the world. Her new mainsail had just been delivered, weighing over a ton, and the crew was in need of extra hands. To thank me for my efforts, the skipper invited me to sail with them the next day. I never ended up taking a cabin on that sewage barge, but I crewed *Velsheda* all that summer and most of the next, until an errant wind carried the salt-tang of adventure from T.S. *Astrid*, a magnificent square-rigged brig berthed nearby. Edging along the yard arm to furl the upper t'gallants in a gathering storm soon became my favourite thrill, and once I'd completed my studies, I cast off for new horizons, navigating open waters in the Caribbean.

The pursuit of discovery has always driven explorers. The ships of Christopher Columbus with their indigo-dyed hemp sails embodied the hopes of a Europe on the brink of transformation. It was a time of geographical investigation when everything previously believed was challenged and re-examined. Although most by now agreed that the world was round, some still doubted whether there would be sufficient wind to power a ship uphill if they sailed to the nether regions of the globe. Yet those who persevered against the unpredictability of the winds, who believed that any point on the globe could be reached by travelling in either direction, were the ones who succeeded in reaching their destination – often with devastating consequences. And as new paths of exploration opened up, so too did new ways of extracting wealth from nature.

Today, we are once again confronted with these same challenges of how we share and extract from our small planet and beyond. Whether it's mining space or the deep sea, we face important questions over the enclosure of nature and the privatisation and extraction of wealth from shared commons. What's at stake now, as it was in 1492, is the very shape of the world to come. But why should it only be the wealthy and powerful few, driven by greed and ambitions of glory, who change the course of history? Why not also the many, propelled by a creative vision of how things could be: the rise of the commons, rather than the inflated profits of corporations who carve up and privatise our shared spaces? I'm reminded of the red-sailed luggers, favoured by Devon smugglers of the past, which could easily outmanoeuvre the revenue cutters that patrolled the bay. Sometimes the small and agile can outwit the mighty with a little ingenuity and a willingness to adjust our sails to erratic winds, guided by our internal compasses and an eye to external forces, dodging squalls and skimming immovable rocks that dare block our way.

And while, like dead-reckoning sailors, we need to look back to see where we've come from, trimming our sails to increase speed and avoid errors of magnitude, we must also keep our eyes on the horizon. Out beyond where the sea mist lingers, obscuring our vision of what is always there, we might just catch a glimmer of the richest treasure of all: a land of hope; a state of renewal. I believe it exists. We just need to find it. Of course, a ship is only as good as its crew, so let's sail the open seas together, a flotilla of diverse efforts coming together from every corner of the world. There are infinite unwritten futures to choose from. All it takes is the courage to steer into the unknown, trusting the world will reshape itself as we go.

## A lonely island

***Brazil • 1507***

Winds and currents, favourable or otherwise, shape a sailing voyage in much the same way as our beliefs, values and assumptions have altered the course of history. Columbus, despite his certainty, never did find his Indies and, as we shall see in this story, Portuguese navigator Pedro Álvares Cabral's voyage to India would be rewritten by unforeseen drifts. Yet as much as these early encounters were framed as accidents, they reveal a deeper truth: our tendency to interpret the world through the lens of our own expectations means we can fail to see what is truly there.

Pliny the Elder, in his *Natural History*, was wrong about the source of indigo. Marco Polo mistook rhinoceroses for unicorns, and when Cabral's fleet stumbled upon the vast landmass of Brazil, they thought it was an island. These misinterpretations serve to remind us that knowledge is shaped by the limitations of our minds. But what if we embrace those moments when we are thrown off course? What if instead of rigidly adhering to what we think we know, we allow space for adaptation and the discovery of new ways of understanding our world? If those years sailing classic boats have taught me one thing, it's that mastery comes not from controlling the elements, but from learning how to work with them, how to lean into the wind in order to pick up speed, adjusting sails or course whenever necessary.

---

The fleet set sail from Lisbon with a clear purpose: to follow Vasco de Gama's route around Africa to establish Portuguese trading posts on the spice-rich shores of India. Borne westward by favourable winds, the expedition sailed a little too far. It was to prove a fortuitous error, for Portugal at least, for one day the lookout spotted a hump in the ocean that wasn't a whale. Cabral turned his thirteen ships towards it; he may not have been the best navigator in the world, but he was certain this couldn't be India. Whatever it was, it lay on the Portuguese side of the line traced by the Pope six years earlier, dividing the New World into spheres of influence. An island to claim for the King was reason enough for a detour. With fresh wind in his sails, it wasn't long before Cabral's silver-buckled feet were firmly planted on terra firma. Island of the True Cross, he named the place, tarrying just long enough to string tin crosses around the necks of the curious sightseers, men and women who went about naked save the red and black patterns that covered their bodies. Dispatching one of his vessels to take news to the King, Cabral set off for the East Indies once more. All he left on the island were two seasick convicts whose complaints had worn thin and a cross in the sand.

Back in Lisbon, the King loses little time in dispatching an ambitious young Florentine, Amerigo Vespucci, to chart the new territories. The Island of the True Cross is swiftly renamed the Land of the Holy Cross. This is no island, confirms Vespucci on his return, because the coastline is long and unbending with countless inhabitants. Nor is it part of Asia. *I have found a continent in that southern part, more populous and more full of animals than our*

*Europe, or Asia, or Africa, and even more temperate and pleasant,* he contends. *It is lawful to call it a new world, because none of these countries were known to our ancestors.*[4]

Vespucci confirms the existence of an immense quantity of gold lying under the ground, not to mention all the riches contained in the forests. Because the continent is new and no one knows what to call it, a young German cartographer tentatively suggests they might consider naming it after the man who uncovered this unmapped world; after all, he reasons, both Europe and Asia are named after women, and some claim the word Africa harks back to the original motherland. He draws a fourth continent rising from uncharted waters, its position based on speculation and imagination as much as astronomical observation. America is the name he writes across the landmass that looks like a long stocking on a thin bony leg.

The King grants a lease on the Land of the Holy Cross to a monopoly of merchants who immediately set off to explore, to expand, to extract, with a percentage of profits to be paid to the Crown. After charting broad swathes of the coastline, the venturers return to Lisbon with a cargo of wood. This, they tell the King, is what we will trade. The locals call it *ibiritanga* – red tree – and from it comes a dye that flames like a hot coal on the fire, *como uma brasa no fogo*. This tree is already known to us. It is the very same species as sappanwood, the one some call *pau brasil*, for which we pay scandalous prices from the east. Back across the ocean they sail and start cutting down trees, making a great fortune for themselves and their King in the process, for all over Europe there are few emblems of affluence more vocal than the chromatic intensity of crimson. The trees grow in such abundance, the merchants dub the place Terra do Brasil, Land of Brazilwood, and the name sticks, although there are one or two unquiet consciences at the Holy Cross being traded in for commerce. To support the merchants' monopoly, the Crown prohibits the import of any competing dyewood from Asia. The Portuguese have red all sewn up – or so they think.

But so vast is the coastline that the French go there too, sailing in and out almost undetected, looking for their own riches to trade. The weather is clammy, so one day a crew member volunteers to wash the men's shirts. Following the normal procedure, he boils up fresh ash from the fire, mixes it with lye soap and sets about scrubbing the collars and cuffs. To his surprise, instead of whitening, the shirts turn bright red! No amount of soaping will get rid of the colour and the crew are forced to wear them that way. They don't stay disgruntled for long. Once it dawns on the men that the wood ash is a red dye, and an immovable one at that, they know their fortunes are assured.

Fifty years: that's all it takes for thousands of brazilwoods, the giants of the Atlantic seaboard, to be reduced to stumps. But shifting winds of fortune hardly trouble the competing crowns who are much more interested in the new commodity on everyone's lips – a new reason to cut down trees. This time they'll be clearing land for plantations, because sugarcane is the future, so they say.

---

Cabral's unintentional detour had devastating consequences, not just for the land that would become Brazil, but for its people, culture and natural world. His arrival marked the start of centuries of colonisation, exploitation and the erasure of entire civilisations in the name of expansion and profit. These errors were far from benign: the exploitation of brazilwood was a prelude to redrawing an entire continent's future, a blood-red reminder of the dark side of discovery.

Yet there remains a broader truth to be drawn from Cabral's journey, not about conquest, but about how we respond to the unexpected, because we are always in the process of learning, through trial and error or the vagaries of the winds of fortune. Whether we are charting continents we presume to be new or innovating solutions to the existential problems of our day, history shows that we very rarely get it right the first time. And when we feel we've gone astray, it can be difficult to summon the resolve to pick ourselves up, dust off our tarnished courage, and start all over again. Yet it is precisely in these moments of misdirection and uncertainty that new opportunities can emerge – if we approach the unknown with humility and an openness to creatively rethink our course.

Whenever I speak to young people, I encourage them to set out with questions rather than answers, to search for unexplored connections and to seek out better ways of doing things. To challenge prevailing wisdom and refuse to accept that something can't be done in a certain way until they've tried. The course of history can be shaped by those who are willing to question, adapt and disentangle today's challenges. Or it can be written by those, like Martin Waldseemüller, the young German cartographer, who dared to put pen to paper and say this is how it will be.

We cannot afford to wait for the perfect wind, nor should we rely on the old maps that led us here. Instead, we can approach our own journeyings with vision and versatility, navigating with ingenuity, questioning our assumptions as we go. If we do so, we too may find solutions waiting for us where we least expect them – and begin to reimagine what is possible.

## Possession

*Peru • 1532*

For most of human history, textile production has been woven into the fabric of domestic life. Indeed, up until the late fifteenth century, the word trade meant no more than a path or track, or the course of a ship, the new definition blowing in on the trade winds. Not long after Columbus's voyage, textiles were swept into the current of global commerce as fibres and dyes cross-stitched the world in an ever-widening pattern. Or perhaps it was the other way around, for wasn't it textiles that ultimately transformed the global economy, catalysing technological progress, propelling the Industrial Revolution and increasing consumer demand for all types of products? Whichever way we choose to see it, companies were not slow to cotton on to the fact that where there were cheap materials, there were even cheaper hands. Colonial expansion reshaped not just industries but the fate of nations and the language of ownership itself.

---

Francisco Pizarro arrives in the north of Peru. He has only a few dozen horses and a small force of men, but the horses snort fire and the men gleam with armour like a flotilla of silver turtles. More than that, they are artful and will stop at nothing on their route to conquest. As they march across the dry coastal plain, Pizarro is surprised to come across cotton plants in a profusion of colours. He only knows of white cotton, but here it is brown, lilac, beige, maroon, orange and green, and he wonders at the strange customs of the locals who must have dyed these fibres and set them back on the bushes to dry. He soon forgets about the cotton, however, confident of greater riches ahead.

Atahualpa, the Inca ruler, commands an army of at least fifty thousand and sees no threat in meeting Pizarro, who arrives with only a few hundred steely men. But the Spanish coerce Atahualpa with lies and kidnap him during a peaceful meeting, demanding a ransom of silver and gold. This puzzles the Inca officials, for aside from the storehouses filled with potatoes, maize and quinoa to ensure no one goes hungry during times of scarcity, textiles remain the only form of accumulation in the empire. Atahualpa issues orders to comply with the visitors' unusual demands, but once the bounty has been delivered, they execute the Inca ruler anyway. Little by little, through treachery and weaponry, the Spanish take possession of the realm.

The Spanish take possession of the people too. Why not? they argue. The locals are indolent, uncivilised, incapable of learning or reason – at least that's what the invaders choose to believe, conveniently ignoring all evidence to the contrary. Hands are needed to sow the cotton, pick the indigo, mine the silver and gold, the extraction of resources being one of the rewards of conquest: eighty per cent for the colonist, the rest for the Crown. This presents a problem for Queen Isabella of Spain, who has already banned the enslavement of the region's inhabitants. So, to avoid going back on her word, she gives royal approval for the practice to continue under a more palatable name. The encomienda system, from the Spanish verb meaning to entrust, is framed as a benevolent arrangement in which settlers are entrusted with the care and conversion of Indigenous Peoples. Before long, entire communities will be compelled to give up their language and labour in exchange for the promised salvation of their souls.

There is no word for possession in Runasimi, the language of these Andean people; no mine or yours, no rich man's estate, no poor man's hut. In their native tongue, you cannot own land or trees, or the rivers of gold. You cannot own the cotton and indigo, which grow so well on their fertile soils. And you certainly cannot own people. The *runa*, the people, borrow from the earth and give back to the earth, keeping everything well-balanced in the dynamic give and take they call *ayni*. A flow in one direction is always balanced by a return the opposite way, like the tides. The cotton plant is part of this cycle: it grows, blooms and releases its seeds, which find fertile soil in which the next generation will flourish. And so the flow of life continues.

But there is a word for possession in Spanish – several, in fact. And they don't behave like the cotton plant. Possession is like the purple butterwort, those greasy little flowers of the high Andes that trap small creatures with innocent blooms before dissolving their prey. Once things get inside it, they never come out.

---

Possession is an enticing word, a powerful word – one that can easily lead us astray. As we move forwards to a more collective future, I believe we will need to rediscover a whole lost ecology of thought, a place where words like possess struggle to find their place unless in expressions of shared joy and wonder. If I were to create a Dictionary for the Planet, language would evolve an entirely new kind of conjugation, not for voice, mood or tense, but for collaboration: a grammar of solidarity. This language would replace verbs of ownership with those that express the many ways we can work and grow together, becoming rather than being. It would cease to define what we could take, but there would be thousands of pages listing words for all we could make together. In numberless pages spanning every tongue, we would find a language rich with ayni. And this mutual exchange would set in motion an unstoppable flow of creative ideas, threading an entirely new lexicon of belonging. Just imagine.

## Truth or fiction?

***Brazil • 1546***

Forty years after Brazil fell under the Portuguese sphere of influence, the lure of riches drove European explorers ever deeper into the heart of the land. Now, the focus had shifted from the tree-shorn coast to the enigmatic interior, where the mighty Amazon River streamed a fast-flowing ribbon through the unknown. For Spanish soldier Francisco de Orellana, the search for wealth and glory not only led him down treacherous tributaries, it took him into the verdant heart of the forest, chasing golden cities and encountering powerful women along the way. While Orellana's tales of great cities were dismissed as the fevered dreams of a man lost too long in the jungle, recent science offers tantalising clues that such places may not have been fantasies after all.

---

Francisco de Orellana has heard rumours of gold – not that this is unusual in a land where even fishing hooks glimmer beneath the surface like sunlit treasures. A seasoned conquistador, he has already served under Francisco Pizarro in the bloody conquest of the Incas. His eyes will never forget the storerooms of treasure handed over for Atahualpa's ransom. But this is more, bigger, an entire golden city lying somewhere in the heart of the land. So, he sets off in search of riches that can scarce be imagined.

Led by Gonzalo Pizarro, Francisco's brother, Orellana's expedition descends from the mountains of Quito, sailing down the chocolate waters of the Putumayo until they reach even darker waters that they name the Rio Negro. Dogged by disaster, the company find little along the riverbank but angry insects and even angrier people. 'Go on ahead with fifty-six men,' Gonzalo tells Orellana. 'Scout the land, and take the friar along to record progress.' He accepts without hesitation. A chance to go further, to chase legends, to win glory? Weeks later, when his half-starved men find a village that feeds them, they refuse to return. Orellana seizes the opportunity to continue downriver. He will not stop nor turn around when there's a kingdom to plunder, especially with the justification of mutiny beneath his gold-buckled belt. But to continue, they will need a second boat.

The next settlement they come across offers supplies: timber, food, cotton. The villagers are willing, and within a month, a second brigantine is ready, its decks calked with cotton, the sails cotton too. Off they set once again, fighting rapids, hunger, warriors who appear on the banks painted black with genipap, red with achiote. And all the time, the rumble of distant drums calling the country to arms. The friar faithfully records all these events as they occur, despite the arrow that goes straight through one eye. Whenever the party are able to make it ashore, they trade or they raid. Orellana handles pottery more beautiful than anything he has seen in Europe, finds storehouses filled with cotton, grains, fruit, nuts, and marvels at the turtle farms.

Before long, they encounter warrior women: tall, powerful, naked apart from a small cotton cloth around their hips, their long hair braided and wrapped around their heads. These women fight fiercely, shooting arrows as sharp as porcupine quills; tipped with a liquor that numbs the nerves slowly, their poison shuts down every part of the body without touching the heart. A furious battle ensues. Amazons, Orellana calls them as he counts his losses, after the mythic warriors described by Herodotus and Homer, for how could such creatures be made of flesh and blood? Compelling and terrifying all at once, these figures are the stuff of legend, crossing continents and millennia, and a sure sign they are nearing El Dorado. Yet, the legendary gold continues to elude him. The expedition spares little thought for plants, and the rangy shrubs with spiky red fruit go unremarked. Had Orellana known that *uru'ku*'s crimson stain would one day tint everything from cheese to textiles, he might have looked twice. Or had he realised that some, like the Abá, believed the plant could guide them to a place of abundance, he might have wondered if he'd been searching for the wrong kingdom all along.

After more than a year running the river's gauntlet, the ragged crew reach the sea. Dismayed but defiant, Orellana names the great river for the fearsome women who line its banks. River of the Amazons, he calls it. Most dismiss the existence of such figures, claiming they are as mythical as the legends from which their name derives. No one – at least no man – wants to admit that somewhere out there is a land where strong women thrive, running matriarchal societies as powerful as, perhaps more so than, their own. Orellana returns to Spain to recount his peregrinations, and that's when the new rumours start: of a man so deluded he speaks of millions of people in gleaming white cities with warrior women who fight like ten men. The Spanish Crown regrets it can offer no official support for another voyage. Undeterred, Orellana charts his own course, assembling a small fleet of ships crewed by three hundred men, along with two dozen horses and his fourteen-year-old bride. He doesn't get far.

Orellana dies at the mouth of the river he has named. Some say he drowns, others that his body was found riddled with the arrows that kill with a whisper. But his name will live on, less for his exploits than for an evergreen shrub whose prickly fruit contains glossy red seeds – the plant the people of the forest call uru'ku. Although Orellana himself showed no interest in the shrub, his expedition – the first European voyage down the

Amazon – left behind a detailed account of the land and its flora. More than two centuries later, when Swedish botanist Carl Linnaeus was devising his system of plant classification, he chose to commemorate the journey that introduced Europeans to the rainforest by naming one of its plants after the expedition's leader: *Bixa orellana*. Yes, Francisco de Orellana's name is preserved, not in gold or in glory, but in a small seed that stains your hands the colour of blood.

---

The Amazon keeps its secrets well. For nearly five hundred years, stories of bustling cities along the river were dismissed as the ravings of a man lost to his delusions. Yet recent discoveries of *terra preta* – dark, fertile soils enriched by human activity – suggest the settlements described in Friar Gaspar de Carvajal's chronicles may have been real after all. One study uncovered evidence of over ten thousand pre-Columbian earthworks throughout Amazonia, while other researchers speculate the region may have supported up to ten million inhabitants. While most soils in this area are nutrient-poor, these carbon-rich deposits hint at sophisticated societies employing practices of soil enrichment that continue in Amazonian communities to this day. Not only did this fecund earth nourish crops, but it serves as a carbon sink – a gift from the past to sustain our present.

As we dig deeper into the mysteries of the Amazon, it becomes increasingly clear – especially to those of us whose worldview is filtered through a pragmatic Western lens – that the line between legend and reality is not always as sharp as we think. Orellana may never have found the golden city he sought, but the radiant seeds Linnaeus named *Bixa orellana* have been part of the Amazon's story for centuries, brilliant symbols of the land's true wealth; riches measured not in gold or silver, but in the ecological wisdom that sustained diverse cultures long before European boots tramped the shore.

Today, as the world looks to the Amazon and its peoples for both resources and solutions, I am reminded that true wealth lies not in what can be taken, but in what is left undisturbed, both in the ground and in the spirit of reciprocity that endures between Indigenous Peoples, their lands and the natural treasures they have nurtured for millennia.

## New fires

***Mexico • 1562***

History often unfolds quietly, with those living it unaware of their place in the grand sweep of time. In the years before this story takes place, a Mexican scribe painted an account of the New Fire Ceremony, a familiar cycle marking the resetting of time in a world woven from sacred patterns. Every fifty-two years, a priest would knot a maguey rope around a bundle of reeds, symbolically binding up the people's long yesterdays before kindling a new sacred flame that would light their tomorrows. A new fire to carry them into the future. Among the deities depicted in the scribe's guide was Mayahuel, goddess of maguey – the plant we know as agave.

For the Mēxihcah – soon to be renamed Aztecs – the maguey plant was *metl*, a life-giver that took countless forms. Some varieties were named for their appearance: *metlcoztli* with yellow-edged leaves, *huitzitzilmetl* with a red tinge to its roots and thorns and *nexmetl*, the colour of ashes. Others were prized for their properties: *teometl* cured fevers, *tlacametl* restored a woman's strength, while the roasted leaves of *mezcalmetl* were a delicacy. But for Mayahuel's portrait, only one variety sufficed and that was *quetzalichtli*, source of fine thread; named for her lover, Quetzalcoatl, it was a rare one without metl in its name. Rising from maguey's spiky leaves, Mayahuel holds a coiled maguey rope in her hand to signify strength and creativity and is crowned with a headdress of spindles, drawing maguey into thread, spinning chaos into order. As the scribe painted his guide to the New Fire Ceremony, the cycles of the universe must have felt immutable, as if nothing could disrupt their sacred rhythms.

But the future often arrives unannounced. Away to the east, the shadows were gathering, ready to unravel the familiar threads of his land, his gods and their cycles of fifty-two years. While the scribe was putting the finishing touches to his manuscript, a German mapmaker picked up his pen and gave America a new name. Before long Tenochtitlan, the mighty capital of the Mēxihcah, would fall to the Spanish, its temples toppled, its deities disdained. And a whole new era would begin.

---

There are two kinds of priest in the Americas. When they depart from Europe, they are on the same mission, but one type of priest will be changed by what they see. Fray Bartolomé de las Casas is the first kind of priest, not that he was a priest at all when he first arrived on Hispaniola at the age of eighteen, but the son of a merchant who accompanied Columbus on his second voyage. Like the rest of the settlers, Bartolomé received his due portion of land and slaves. And an education in extermination. The Spanish were, he recounts, like starved lions set loose among innocent sheep: hacking, slicing, drowning, hunting Indians for sport, not sparing the old nor the young.[5] It was all too much for young Bartolomé, who returned to Spain to become ordained as a priest. But then he went back, following his calling; back to Hispaniola to coax the locals to the faith, or at least those who remained.

But his own conversion was not yet complete; that wouldn't happen until one Christmas Eve morning when he joined the congregation for Friar Antonio de Montesinos' sermon. Expecting the familiar Christmas story, he settled himself into the hard wooden pew, the cathedral's thatched roof providing welcome respite from the searing heat. All the worthies of Hispaniola were there. But the friar had something else in mind that day, taking the opportunity of a packed church to launch a blistering attack on the colonists. Bartolomé had an epiphany, renounced his slaves and vowed to dedicate his life to upholding the rights of the oppressed.

Even when he becomes Bishop of Chiapas, deep in Mexico's south, Bartolomé never ceases questioning by what moral right the colonists conquer and enslave. These people are not idolators, nor are they barbarians, he avows. Indeed, they surpass the Romans, Greeks and Spanish when it comes to culture and customs. That being so, he reasons, if we are to follow Aristotle's order of nature, Indigenous Peoples should be the natural rulers of the Spaniards. Yet, even as he highlights these differences, he affirms a deeper truth: that we are part of a common humanity. Four hundred years before the UN Declaration of Human Rights, Fray Bartolomé de las Casas concludes that the entire human race is one.

Fray Bartolomé, Protector of the Indians, former Bishop of Chiapas, has spent fifty years in the Americas. He is nearing eighty now, but the fire in his belly is undimmed. He puts quill to paper and writes to the Council of the Indies, the governing body for all Spain's overseas realms, begging the powerful gentlemen to listen to what he has to say. With little to lose, he denounces the conquest as tyranny, the encomiendas too. And what of all those riches pouring into Spain? Very little has been acquired by legitimate means; the vast majority has been stolen. And by 'acquired' he alludes to gold and silver from islands and territories where everyone had been killed by the Spanish or wiped out by disease, leaving these lands empty of people to steal from.[6]

While Bartolomé de las Casas is penning his letter, another priest to the north is at work. Diego de Landa belongs to the second group of priests who have come to the Americas; he has been in Mexico for thirteen years, enslaving, torturing, extracting confessions, unmoved by what he has seen. Under his watchful eye, the amate paper crackles and sparks, and Mayahuel flares into light. But this is not another New Fire Ceremony. The text over which the scribe laboured long is spitting on a bonfire, its flames stoked by pious zeal. Two boys have found a cave full of idols and evidence of sacrifice – perhaps bloodied maguey thorns even litter the floor – and it's time to turn up the heat. Now that Diego de Landa has torn down the Maya temple and built a church with the stones, he will extirpate their noisy gods in the roar of the flames. This is how Diego de Landa intends to eradicate rival traditions of literacy. Twenty-seven demoniac documents are burned at Mani in the Yucatán, their idolatrous owners put to death. Codices, the Spanish call the papers, filled with coded messages of unpronounceable gods: gods with animal names and numbers; gods with breasts; gods with spindles in their hair; gods who dance. Mayahuel rising out of the maguey plant, its spines sharp as truth. By the end of the night, millennia of knowledge has gone up in smoke. Or at least, that's what he thinks.

Diego de Landa fails to realise that these figures have a life beyond the illuminated manuscripts he has erased. They are imprinted on hearts, remembered in dreams, retold through song, revived anew every morning. The Spanish may have put an end to the compulsory education enjoyed by both girls and boys, being at odds with their insistence on the ignorance of the local population, but they cannot stop these stories weaving through minds. Diego de Landa does not understand that whenever a woman carefully strips the leaves of the quetzalichtli maguey, she thinks of the tail feathers of the quetzal bird. And as she spins its fine thread, she turns her face and sees Quetzalcoatl: inventor of books, protector of artisans, lover of Mayahuel. Quetzalcoatl, giver of life. Once her maguey is spun, the woman will begin making loops for her bag, counting them off in the Mayan base unit of twenty, because why create a decimal system when you have ten fingers *and* ten toes? Then she will add adornment, passing a strand through a needle broken from agave's protective teeth.

For Diego de Landa, it is only texts that are dangerous. He does not recognise that weaving is more than a means of making a textile. And so, up and down the land, women work tirelessly, drawing sustenance from Mayahuel, threading resistance into every last stitch.

---

There is a saying that tradition is not the worship of ashes, but the preservation of the fire. Long after the codices had gone up in smoke, the flames of resistance continued to blaze – a living fire stoked by shared stories. Far to the south, Bartolomé de las Casas, ageing and frail, did not live to see this future, but his gentle beliefs in social justice and a shared humanity would outlast the sharp practices of the Spanish oppressors. For while a bonfire burns bright, it burns fast; before long, its grey ashes become one with the wind. But the patient work of those who remember, who weave the threads of their past into the future – this is what endures. Not just in the textiles themselves, but in the expression of living knowledge, imprinted on minds, remembered by hands, out of reach of those who seek to destroy.

## Stamping down yellow

***Ireland • 1570***

As the wind fills our sails and carries us across the Atlantic, we hear another struggle rising, not over codices, but the colour of clothes. For Ireland too was contested ground where the heavy hand of colonial power sought to quash the spirit of the poor, and where cultural identity solidified under oppression.

---

Galway was an outpost of England, a city with a wall around to keep the wild Irish out. It was a confident, preening place where none but the wealthy could claim citizenship. Galway, its privileged inhabitants boast, rivals Dublin when it comes to fashion. Woollen doublets, silk gowns, embroidered cloaks and wide pleated collars go in and out of stately homes and prosperous merchant houses. Nothing too ostentatious of course, because Protestantism is all about moderation, and they want none of those flamboyant colours worn by the poor. Sad hues are in vogue and, in any case, colour is regulated by the Crown.

Galway has only one problem: what to do with the riffraff? And they're hard to miss in those bright yellow tunics. The poor are all Catholics, of course, and Gaelic-speakers too. But why must they all wear the same filthy uniform? Their tunics are still dyed with weld, but in recent years another yellow has come on the scene, and this one's even brighter, if that's possible.

This yellow is saffron, true saffron from Spain. There is only one problem for the good Protestant citizens of Galway: saffron has a stigma. Saffron is yellow, yellow is popish and, ever since the powdery filaments of the small crocus flower turned up on their shores, it has rivalled weld for Catholic affections. In fact, the luminous hue has proved so popular that saffron now makes up one-fifth of their imports. Not that they mind the trade, of course, but they'd rather it was a dye untainted by poverty, popery and dirt. And now they have an ally in the English, who have taken against the colour too. In fact, saffron has become a symbol of everything that is abhorrent to the emergent Protestant nation, who find any excuse to express their distaste. Saffron is papist, effeminate, traitorous, worn by brain-sick men and painted women. It is used to prevent lice. It uses urine as a mordant (not that this is uncommon in dyebaths on both sides of the Irish Sea). It is wasteful to use so expensive a commodity on peasant dress; aren't there better ways to spend their hard-earned groats? And on they grumble.

When vilification doesn't work, the Crown tries to ban the colour instead. A letter arrives in Galway from Henry VIII, desiring *no man, woman, or child do wear in their shirts or smocks, or any other garments, no saffron, nor have any more cloth in their shirts or smocks, but 5 standard ells of that country cloth.*[7] The poor can't understand it, no matter how many times the letter is read out to them in the marketplace. Five ells, when they were used to gathering thirty or forty into their pleats? Does the King want them walking about half-naked? They cannot possibly adhere to one half of the law, and have no intention of complying with the other, so potent is the colour's connection with their identity. In any case, who can tell whether their clothes are dyed with the real stuff, or with poplar leaves, wild arbutus, weld, heather tops, gorse flowers or any of the other native saffron hues? The King writes again, repeating his demand, and in case anyone doubts his seriousness, he extends the regulations to cover the entire country. The *Act for the English Order, Habit and Language* decrees that no shirts, smocks, kerchiefs, neckerchiefs or linen caps can be coloured with saffron, nor can anyone wear above seven ells of cloth, nor add any silk trims or embroideries – not that there's a chance of such fineries when you live off the land – nor wear any mantles, coats or hoods in the Irish fashion. No one heeds this new law, not around Galway at least; two more measly ells barely widens a cuff. In the end, the local authorities shrug their shoulders and impose a two-pence a pound tax on the yellow stuff instead, using the profits to enclose common ground because turning grazing into genteel parkland is at least one way to keep the gentry and rabble apart.

In the end, it's not the law that kills yellow, but fashion. Exotic dyes continue arriving in Galway's bustling harbour: brazilwood, logwood, kermes, orchil, verdigris and indigo. The temptation to experiment proves hard to resist and a whole new sartorial colour palette is born.

---

The battle over saffron was never just about colour; it was about what yellow represented – a beacon of faith, language and belonging that refused to dim under foreign rule. Just as the cultural memories of the Mēxihcah and Maya survived the burning of the codices, so the values held by Irish Catholics could not be legislated out of existence, however hard the English tried. And it makes me wonder whether the seeds of silent resistance, planted in the soil of shared history, put down the deepest roots.

## Bothersome blue

***England • 1600***

In England's evolving landscape, colour was rarely employed for aesthetic purposes alone: it was an emblem of politics and power, tangled up in the weighty cords of commerce, religion and empire. Yellow was not the only hue bothering the English. Blue was proving troublesome for an entirely different reason and, once again, it was the combination of colour and Catholicism that sparked fury.

---

England has a problem with blue. First came those hard blocks of indigo, long mistaken for a mineral, but the pigment known as Indian stone proved a storm in the dye pot as none could afford it save the wealthy and the artists who painted them. Next it was logwood, shipped to English shores by the Spanish. The small tree yielding this new blue grew abundantly along Mexico's Yucatán coast. So plentiful was it that the Spaniards used the wood as ballast, cutting it into three-foot logs to stabilise their boats on the long voyage home – at least they did until they realised it was a dye. Logwood was cheap and it wasn't long before broadcloths, wools, cottons and flannels began turning blue. Some even used it to make black, a colour becoming increasingly fashionable in life and in death, but difficult to achieve without countless alternating dippings in madder and woad. The woad growers, dyers and merchants were furious; they weren't against the democratisation of blue as long as it came at their profit, not at their cost. Logwood isn't lightfast! they cried. Everyone knows those perfidious, profiteering Catholics can't be trusted. Surely, the Queen's loving subjects should be protected from such artless deception? And look what an industry woad is turning out to be, providing seasonal employment for large numbers of the poor.

What they said was true, for the Spanish had imported the dye, not the know-how. The Maya had used logwood for hundreds of years, but they knew it needed an iron mordant to hold up to light. Not only that, they had learnt the dangers of chopping down too many trees. More than six hundred years earlier, the expanding populations of Tikal and Calakmul had also cleared the logwoods, not for dyes but for plaster, without realising this would open their environment to the stresses of drought. Those cities collapsed, their inhabitants migrating in search of lost vegetation.

In England, opposition continues to grow to the fast-fading blue until Parliament is obliged to pass *An Act for abolishing of certain deceitful Stuff used in Dyeing of Cloth*. The growers, dyers and merchants sigh with relief: they've kept the lid on blue, at least for the time being. But laws are little good unless they're implemented. Logwood still creeps in under shade of night, with scarcity inflating the cost. The time has come for England to draw a secret weapon from up her lace-cuffed sleeve: pirates – or privateers as Queen Elizabeth prefers to call them. Sir Walter Raleigh and other gentlemen of his ilk were adept at relieving foreign ships of their cargo, including these valuable dyestuffs. Tensions escalate, culminating in the sinking of the Spanish Armada. This leaves the English with a straight run at logwood. And as the imports of privateers make a mockery of the law, down come the prices again.

Up go the cries of the woad growers, but this time no one is listening; in fact, they're starting to feel the whole country has turned against them, from the soil to the Sovereign herself. It began down in Hampshire, one of the best woad-growing regions, when the Corporation of Southampton decreed that no one should break ground for the purpose of sowing or planting woad, finding neither grass nor crops would grow after the nitrogen-greedy plant had been sown.[8] Like a spilt dye vat, streams of discontent flow through the land. At Ely, where woad has been grown for little more than a year, the parish requests its sowing to cease for fear of increasing the number of paupers once the soil is exhausted. Little by little, England rallies against woad. The land must be used for grain to feed hungry people, they say, and that's when the Queen herself issues a proclamation against the sowing of woad, citing the inordinate gains made by the few at the manifest cost of the masses.[9] No person is to sow woad within four miles of a market or clothing town, or any city, or within eight miles of any house of the Queen. Yet, like the logwood law, many choose to ignore it. So many in fact that woad sales increase, and that's when Her Majesty decides to ban the breaking of new ground to sow woad across the entirety of the realm.

And who does this benefit? cry the woad growers, merchants and dyers as woad becomes the second-biggest import from France after wine. They petition, they lobby, they make their views heard, until Her Majesty is forced to back down. A new proclamation allows anyone anywhere to sow woad on their land, the only restriction being around cities and royal residences due to the miasma of cabbage that drifts from the dye vats. It's as close to an apology as you'll get from a Queen. Finally, the growers, merchants and dyers see clear blue waters ahead. Little do they know it's only a lull.

---

As the controversy over woad started to fade, England believed it had found peace in the control of colour and commerce, although in reality the battle over blue was far from over. This tension between what is visible and controlled, and what escapes notice and is therefore ignored, like the natural fibres persisting in aquatic environments, reflects a broader truth, one we are coming to recognise time and again through these chapters. Power lies not only in what is seen, but in what is overlooked.

## Seeing differently

***Peru • 1615***

What would it be like to see the world differently? As I write this chapter in my Branscombe caravan, perched at the confluence of worlds and eras, I gaze out at the sweep of October sea and try to picture the world as my forebears who lived here would have seen it. What would it be like not to know that out to the west, beyond Berry Head where a lighthouse flashes, lay an entire continent? While my ancestors were tethered to this patch of land through the centuries, attuned to its myths, rhythms and seasonal cycles, their lives found an unheard echo across the ocean. Perhaps back then we were not that dissimilar after all? Yet today I am struck by the difference between Western and Indigenous worldviews, far vaster than the short space of time and ocean between us, and I try to imagine what it would be like to see the world through new eyes.

I glimpsed something of this in my masters in Native American studies when I wrote a paper analysing Felipe Guamán Poma de Ayala's drawings in his 1615 document *The First New Chronicle and Good Government*. At first glance, the illustrations might appear hasty, naïve, mere adornment to the weighty text. But this is far from the truth, for within those drawings lies hidden meaning, a judgement of the conquest and its aftermath that uses spatial positions to communicate Andean concepts of power. There was only one problem with carrying out this research. Guamán Poma's illustrations are drawn as if you were inside the page looking out, like a director backstage giving instructions to actors. Right is left, left is right. And positioning is crucial. That Inca warrior may appear to have been defeated by a conquistador, but when you understand the values encoded in the drawing, they reveal his moral superiority. As I said at the end of the previous story, power often lies in what is overlooked.

For months, I forced myself to change my reality, not in an abstract philosophical sense, but in a very literal one. I immersed myself in Guamán Poma's way of seeing the world, a place where space is not linear and relationships play out across planes unfamiliar to Western eyes. After I handed in my paper to my tutor, Dr Valerie Fraser, I found it hard to return to my old ways of seeing. For months, I muddled my left and right, like borrowing a car with indicators on the opposite side, and when you return to your own you fumble towards junctions wiping the windscreen. It was disconcertingly hard to readjust. But it made me realise that the patterns of our mind might not be as fixed as we think. Perhaps changing our way of seeing – and by extension how we act in this world – is not as impossible as it seems.

---

Felipe Guamán Poma de Ayala does not see the world as the new King of Spain sees it, so he decides to send him a letter, suggesting the King might like to take a leaf out of the Incas' book. Except the Incas do not have books; they do not put their taxes and tributes, histories and narratives, maps and genealogies down on paper. Instead, they are tied up in string. Every colour, every knot, every placement, every twist tells a story. Their language is not written: it is spoken, it is felt. Khipu is the name of their writing system, but the Spanish haven't bothered to learn the language, so Guamán Poma, heir of the Yarovilca dynasty that predates the Incas, has decided to write it all down.

It is turning into a very long letter: 1,189 pages to be precise, mostly in Spanish, with occasional lapses into Runasimi when he can't find quite the right word. By the time he has finished documenting daily life under the Incas and compared it to existence under colonial rule, Felipe III has been on the throne some fifteen years. To add context to what he has to tell the King about the abuses going on in his name, Guamán Poma illustrates his letter with almost four hundred drawings in his own hand, revealing not only what was being lost, but what the Spanish never saw. And while his written discourse is rhetorical, designed to persuade by setting out the facts in what he hopes is a reasonable tone, his pictures do more than corroborate the text, at least to those who know how to read them.

Felipe Guamán Poma de Ayala does not see the world as the King of Spain sees it. His drawings encode meaning few white men understand, although they might just wonder why his map shows the west in the east, the north in the south and the sun positioned not on the right, as in European convention, but on the left where the sun of Cusco is born. Guamán Poma draws as if he was looking at the world from the inside out: from the perspective of the drawing itself, not the person looking down at it. In his version of the world, the city of Cusco lies closest to the sun, while Castile and Rome are at the bottom in Antisuyo, an uncivilised place without laws, where people still wear leaves and crude twisted fibres. Because in Peru, it is spinning and weaving that are the markers of civilisation, not silver and gold. Of course, those elements are present here too – the tears of the moon, the sweat of the sun – but their value lies only in art and decoration.

With textiles being the most important commodity in the realm, everyone recognises the drop spindle as a symbol for industry and commerce, so when Guamán Poma comes to illustrate the census categories of women, all except the oldest and youngest are spinning, weaving or collecting plants for the dye pot. And while his drawing of the conquest shows armed Spaniards on horseback facing Atahualpa, the Inca ruler, who wears only cloth, even Guamán Poma's skilled hand cannot convey that the Inca's quilted cotton armour was so densely woven it repelled arrows and spears. Even the outfits worn by messengers and interpreters, Indigenous men with one foot in each world, speak volumes about where their loyalties lie. Textiles are about more than cloth: they are a signal, a gateway, a visual code to past knowledge. More than this, they embody the complementarity at the heart of Andean thought, a communication system for those that have eyes wide enough to see and ears pricked to listen.

Felipe Guamán Poma de Ayala does not see the world as the King of Spain sees it. Everything is drawn according to how his ancestors understood things to be, and whenever he departs from this form, it is a sign that things are not right – a world out of balance. The Spanish overturned the true order of the universe when they came to take possession of Inca lands. Possession: that's another word he has tried and failed to comprehend. The world appears differently through Andean eyes.

No, Felipe Guamán Poma de Ayala does not see the world as the King of Spain sees it. Although it makes little difference. The King never receives his letter.

---

Through my caravan window, a pale sun silvers the foam careening across a steely sea at the tail end of a storm. On the table before me lies a facsimile copy of Guamán Poma's manuscript, acquired at some expense thirty years ago. This monumental letter is far more than a document out of time: it is an indictment of the brutality and injustices that shattered a way of life, and it is a vindication of Andean values that, unlike their silver and gold, could never be stolen. At the heart of the manuscript is a *mappa mundi*, in which the world as we know it is turned on its head. That same map inspired the name of my first brand, Pachacuti, the world upside down. Yet, in many ways the whole book can be read as a map, a chart guiding us towards seeing the world as it once was, and perhaps still might be.

The following day I awaken to a typically autumnal Devon morning. The land blurs into the sea and the sea fades to sky. All clarity is lost to the sea mist tumbling down the cliffs, as if a watercolour artist has blotted the lines of the world. The white horses are gone too, the waves swelling roundly like old grey mares. The rain is coming down. It's not a day for walking to St Winifred's, an ancient church with roots back to the tenth century, so I jump in the car and wind my way through the notoriously narrow village lanes. Today is Branscombe's Harvest Festival and, despite the weather, every pew is full. A red admiral butterfly, drawn by the flowers or the warmth, is rising and falling above the choir in the Norman bell tower. When Father Stephen climbs to the high pulpit for his sermon, he begins with the words: the essence of life is gift. I reel the line over in my head. The essence of life is gift. Simple, clear, without an excess word, it is a statement of fact.

By the time I get home to my caravan, the phrase has lodged itself firmly in my head. And I begin to wonder, if we can open ourselves to seeing differently, to believing and practising that the essence of life is gift, might this not be the foundation for a new way of seeing, entwining the shared threads of our mismatched histories? If the natural world can become other, perhaps, if we find a new perspective, we can recreate the world in its original state, humanity and nature as one.

## A harvest of weeds

***USA • 1619***

For the Englishmen who set sail in the early seventeenth century, colonialism promised a tapestry of opportunity, combining the acquisition of wealth with knowledge of new processes to elevate England's industries. Yet many of these tales were spun from threads of artifice, woven into the public's minds by skilled propagandists. Much like today's distorters of facts, colonialism's architects of deception appeared in various forms.

---

Richard Hakluyt never set foot in Virginia, but his words sailed there. His writings, swollen volumes of patriotic fervour, reached the ears of noblemen, merchants, common folk, even the Queen herself. Don't you see, he writes to a wool merchant, that our English cloth, although the finest in the world, lacks the vibrant colour of those from the Continent. Now imagine mastering the dye techniques of other cultures, allowing us to saturate the world's markets with our bright-coloured goods.[10]

Dyes are just the start. America is brimming with opportunities, gold and silver simply waiting to be unearthed, and with Spanish settlements spread thin, the land is ripe for encroachment. For page after page he spins his grand narratives, blending economic ambition with moral superiority, for this man is skilled at the art of propaganda. His books are filled with carefully edited accounts from those who had actually journeyed beyond the comforts of their London studies, their firsthand knowledge cut and twisted to fit his vision. Across sixteen volumes, Hakluyt conjures a world where wealth could be plucked like ripe fruit, where anyone could rise above humble origins, where new empires enabled them to become masters of their own destiny. He paints a persuasive image, one that colours the minds of the ambitious and desperate alike. In case the Queen should waver, Hakluyt seizes on the account written by Fray Bartolomé de las Casas in which he describes the suffering of innocent lambs at wolfish Spanish hands, hardly needing to add that Queen Elizabeth – for whom the colony of Virginia is, after all, named – would become their benevolent liberator. England's mission would not be one of conquest, he insists, but of prosperity and liberty for all. And so the colonists set sail: noblemen seeking their fortune, merchants looking to expand their trade, younger sons of landowners without an inheritance, tradesmen skilled in their craft. All are drawn by a mirage, a fabrication of bounty just beyond the horizon.

The air around Jamestown is thick with ambition, yet the new settlers are surprised to find themselves surrounded by stubborn earth and wild forests. Yes, the rivers glitter, with false hopes that could drive you mad; minerals sent back to England prove to be base metals rather than gold. These men are not labourers and most, save a handful of convicts, are unwilling to put hand to soil. Desperation is rising like smoke from their campfires. As relations deteriorate with the Chickahominy, who have been supplying them with food, some are reduced to eating shoe leather. But Hakluyt's words ring in their ears. Surely, in the entirety of those sixteen volumes there must rest a few grains of truth? Virginia's lush countryside, thick with strange trees and plants, must harbour something, anything, to sow seeds of prosperity. All they have to do is find it.

The settlers turn to another of Hakluyt's suggestions – dyes. First, they try sumac, or shoemake as some call it, its dried leaves a sought-after preparation for dyeing and tanning, valued by the Cherokee and other Native Americans, as well as by Europeans. The results are patchy at best. They move on to other curiosities: the seeds of the *wasebur* herb, small roots called *chappacor* and the bark of the *tangomockonomindge* tree. They've observed the locals using these plants for dyeing deer skins, rush baskets, mats, as well as faces and hair, but will the colour be fast on linen and cotton? It's a big risk to commercialise dyes without understanding how they work, as the Spanish found with logwood to their cost. No matter, the colour of money takes on different hues. Woad is perhaps a safer option, that plant of ire and vent, for while it is despised in the old country for sapping the soil and its cabbagey smell, Virginia has land enough to spare. Woad could surely be cultivated, madder too, but there's another red that's piqued their curiosity, one with the potential to yield an even greater profit.

*Mutaquesunnauk* is a pleasant fruit, the shape and size of a pear, and a perfect red hue within and without. It's unfortunate the plant has prickles as sharp as any needle, but it's a minor deterrent to their ambitions, for they're certain it is none other than the fabled cochineal. They just don't understand which part yields red dye, as some think the fruit and others its large fleshy leaves; either way, their experiments end in disappointment as the cloth invariably emerges from the dye pot the same colour it goes in. With every new recipe, every failed length of fabric, the promise of fortune slips further out of reach. Some can bear the hunger and humiliation no longer and catch the next boat home.

Running out of options, those who remain shift their focus to fibres. First, they experiment with a wild grass, tall and slender, its gleaming skin peeling away in shimmering strips. They imagine fields of it rippling in the breeze, and a cloth soft as air. But this, too, is an illusion as the fibres are brittle and yields prove too low. Some start to think the land itself is conspiring against them. Then they find hemp growing along the banks of the Potomac River. Cultivated in China for thousands of years, hemp had made its way to Europe, where its strong and durable fibres became indispensable for sailcloth and rigging. Henry VIII even made hemp cultivation mandatory for the nation's farmers and the crop soon earned the name English hemp. Hemp isn't the treasure they had crossed oceans to find, but at least it's a way to give shape to their dreams.

Before long, Jamestown's House of Burgesses, the first legislative body in colonial Virginia, issues a decree. Along with fixing the price of tobacco, prohibiting drunkenness and banning clothes beyond one's social station, it orders every settler to sow at least one hundred hemp plants on their land. As an example, the governor himself pledges to sow five thousand plants of both the 'English' and Indian variety.[11] Without delay, the colony sends for skilled hemp dressers from Sweden and Poland, offering ten shillings as an incentive for any man skilled in the art of separating the coarse woody parts of the plant from its desirable fine fibres. And that's how the age of hemp begins, its threads wending their way into every corner of the colony. The settlers make clothes from it, wipe hands on it, cover wagons with it, blow noses on it, make lamp oil with it. Sailors make sails from its sturdy cloth, stitch the edges with hemp thread and lash them to the spars with strong hempen ropes. Yet still their storerooms are full, so they feed their chickens on it too.

Hemp is making money, hemp is paying taxes, filling in the spaces where gold and silver failed to appear. But for all its utility, hemp will not deliver the easy fortunes the settlers crave. No amount of it could make a man rich, for the cost of local labour remains stubbornly high. To make profit, they need cheaper hands to work the fields. And that's how it begins, entwined with the material life of the colony, binding labour to land in a system of exploitation that deepens with every harvest.

—

Hakluyt's propaganda lured men across oceans, only to leave them stranded in a place of unmet expectations. Today, the echoes of such rhetoric can still be heard, not in the dusty pages of that author's sixteen volumes, but in the polished pledges of modern corporations. During my decade leading Fashion Revolution, I helped design the methodology and conducted research for seven editions of the Fashion Transparency Index. As I pored over sustainability reports, time and again I saw an ocean between promises and reality. Fashion brands spoke of fair wages, yet never defined what a decent salary might look like. They touted sustainable fibres and better materials, but they offered no clarity on what those terms really meant. Empty promises. Hollow words.

All around the world, garment workers are sold a vision of opportunity, of escaping from poverty, often enticed by agents who visit remote villages, promising a girl's dowry upon completion of her contract. Arriving in Dhaka, in Delhi, in Dongguan, these young women find themselves bound to the repetitive capitalist machine, embroidering expensive dresses or stitching cheap T-shirts they will never afford. The promise of good jobs is a fiction, and they're rarely a pathway to freedom. The true cost of material dreams continues to be shouldered by those who work hardest and gain the least. The threads of history are long. And still they bind.

## The soil is our blood

*El Salvador • 1636*

The colonial reshaping of the Americas is often remembered for its brutal conquests and the extraction of precious metals from mines like Potosí in Bolivia, the largest silver deposit in the world. Yet the hunger that drove colonisation ran deeper than the lust for silver and gold, in the same way that conquest was not one decisive slash, but an ongoing act that snipped away at the fabric of everyday life. Over time, land, plants, lives, even entire countries were transformed into commodities. Once bound to their people through cycles of mutual care, lands were tapped like a vein of silver, an artery to nourish distant powers. Such rapacity was more than economic – it was ideological, deep-dyed in a worldview that saw Indigenous life as expendable and nature as something to be mastered.

---

El Salvador, swiftly mapped and measured, becomes a commodity under colonial rule. Land is bought, leased and sold with scant regard for those who have farmed it communally for generations. For the Chorti, Pipil, Maya and Lenca, human life and the natural world are a seamless totality, bound by sacred relationships and spirals of reciprocity. The settlers, fresh off the boats from Spain, dismiss such beliefs as outdated. They boast of their ability to bend nature to their will, converting its bounty into profitable enterprises. Not for them the Indigenous milpa system of corn, beans and squash with its intricate cycles and tangle of synergistic planting – the same traditional method the Cherokee call the Three Sisters. Likewise, the practice of cultivating indigo for two years followed by six years left fallow, which seems absurdly wasteful. The settlers' vision is straightforward: cows, cacao and indigo will comprise their agricultural trinity. While cacao and indigo have long been integral to life in these lands, cows are a recent import, ferried across the ocean as part of the colonists' expansion plans.

Axes come out to clear the forest for cows. The local people try to warn them, but in the settlers' disdain for anything uncultivated, deep ecological knowledge gets buried beneath the weight of ambition. The results are disastrous. The cacao withers and dies in the unforgiving sun, and the Spanish find out the hard way that it is a shade-loving crop, dependent on the canopy of the forest. This is the settlers' first lesson: cows and cacao cannot thrive side by side. Undeterred, the colonial authorities decide to confine cattle-grazing to the east of the country, where Indigenous lands are fair game and, as cows and humans cannot co-exist either, entire communities are pushed out by the encroaching herds. When both cacao and cows fail to live up to their promise, the settlers shift their focus to indigo. This, they believe, will make their fortune.

Of course, the locals know the secrets of the blue herb they call *xiquilite*. They know, too, it is best grown in cycles of rotation to allow the land to recover between planting because while the plant fixes nitrogen in the soil, it still relies on a range of other nutrients for optimal growth. Unhearing or unheeding, the Spanish order continuous planting. All they seek is a fast profit, without understanding the equilibrium that must be maintained for the land to thrive. Exhausted, the soil yields less and less every year, but no matter, there's plenty more uncultivated land to be claimed. The settlers simply shift their plantations a little further along the coast, uprooting villages to scatter new seeds, until indigo plantations stretch the length of the eastern seaboard from Nicaragua to Guatemala.

As cultivation expands, dyeworks mushroom nearby; indigo plants must be processed fast or they lose their pigment. Indigenous workers are coerced into these operations, paid in cheap clothing instead of hard currency to sidestep the Crown's prohibition on forced labour. The dyeworks are a death sentence. Fermenting indigo plants emit a putrid stench that attracts every fly for miles around. When even their beakers of cacao become contaminated, workers quickly succumb to fever, while beneath their feet the soil turns an ever-deepening shade of greed. It's not only the workers who suffer. Forced to abandon their milpas, wives and children fall victim too, severed from the roots that sustain them. Over half the population perishes.

When the land can give no more, the settlers turn their impatient eyes towards people. Not the indolent locals who succumb all too easily. No, this time the hacienda owners fix their gaze overseas. Let us bring five hundred African slaves, they petition the authorities, and when the first ship arrives, they swiftly demand a thousand more. The brutal cycle of exploitation will not only continue, it will deepen, spiralling into something more sinister, more deeply entrenched, for neither land nor life was to stand in the way of indigo's advance. And thus it was that sacred xiquilite, the plant that mapped colour onto the cosmos, evolved into a dark marker of death.

---

The story of indigo cultivation in El Salvador reveals how hunger for profit seeped into the soil, laying the foundation for an economic system that would later be perpetuated by a powerful landowning elite. Guatemala, the Spanish empire's administrative hub in the region, became the centre of the indigo trade, and by the late eighteenth century, around two million pesos worth of dye was being shipped to Europe every year. Meanwhile, it could take indigo workers twenty days to earn a single peso. Yet, their spirit was not extinguished; beneath the parched surface, seeds of resistance had been sown. Centuries later, descendants of those who survived would rise to challenge the landowners. But for now, they bide their time, waiting for their moment to reclaim what was theirs all along.

## Indigo works in mysterious ways

*Mali • 1640*

As dyers struggled to keep pace with the ascension of blue, more and more slaves were packed into ships to work plantations in the colonies. Indigo was a growing concern and slave traders deliberately sought out those with skills and knowledge of this temperamental plant. After the Mali empire collapsed and the Mandé Kingdom embraced Islam, the animist Dogon began a gradual migration across Mali, their roaming eventually leading them to a dramatic red cliff known as the Bandiagara Escarpment. Here, they believed, they would be safe. The Dogon were not the first to settle the place. While its previous inhabitants, the Tellem, were in decline when the Dogon arrived, their presence would have been vividly felt, not just in the remains of their dwellings and artefacts, but through the indigo-dyed textiles hidden in caves.

---

Some call it alchemy, others magic, but however you choose to see it, the art of indigo is intrinsic to Dogon material life. Pull the threads out green from the dye vat and they turn blue as they meet the air, although the exact hue will depend on the type of pot, the quantity of leaves, when they are picked, the source of the water, the number of dips and the benevolence of the spirits. Ever since their ancestors moved to these steep sandstone cliffs, isolation has allowed ample time to perfect the dyer's craft. And these days the Dogon are more grateful than ever for their remoteness from Mali's imperial heart, as it keeps them beyond the reach of the new holy war declared by the Bamana of Djenné, who have risen to oppose Muslim powers and defend their own ways. That's not to say a passing stranger isn't welcome in their villages, wherever they come from because the Dogon have always been lovers of language, and a visitor may turn up with exciting new words. People like nothing better than hearing a new word being born, feeling its sound on their tongue, and they're sure the visitors must enjoy it too, for words grow more powerful the higher you climb. To be without words is to be exposed, vulnerable, like walking around wearing no clothes.

Unlike neighbouring tribes, the Dogon aren't accustomed to buying and selling slaves. On the rare occasions they go raiding, it's usually boys seeking revenge after a neighbour has stolen one of their girls away. While mothers naturally worry for their children's safety, they have learnt to keep their opinions to themselves; no one welcomes conflict in this tight-knit community because good relations are crucial in a place where everyone relies on each other. Of course, there may be an occasional skirmish, but their ancestors were wise to make their new home on a high fissured cliff above a boulder-strewn scree. In any case, if the Dogon do need to hide, there's always the caves where they worship the Nommo and wrap their dead in blue cotton cloth. And if there's no time to reach the caves, any number of men, women and children can squeeze into the hollowed-out heart of a baobab tree.

The baobab's massive presence has watched over generations of Dogon. Growing among their mud-covered homes, millet-roofed granaries and squat earthen towers, its root-like branches cast wandering shadows over the structures below. The Dogon strip layers of bark from these ancient trees once a year to make rope, string and twine, chewing the fibres to soften them before rolling to the desired thickness. The rope and string will be used in their funeral rituals, while the twine is used for baskets or supporting their looms – the looms that make their cloth, the cloth they dye with indigo, the indigo they so stealthily harvest, for knowledge of these plants' cultivation is highly coveted in the lands to the west, where the sun sinks below the horizon on unfamiliar shores. Each time the Dogon strip the baobab's bark, they are reminded of the tree's endurance – a quality they draw on in these testing times as the systems of enslavement practised by their neighbours align with the trade in human bodies shipped overseas. Raids are becoming more frequent too, carried out by the Muslims whose religion the Dogon refuse to adopt. But they can't spend their whole life hiding; they have no option but to travel from time to time if they want to sell their indigo-dyed cloth, so prized by the elite, along with the various crops that they grow, even with the danger that lies below.

The trade routes that criss-cross the Sahara take the Dogon north to Timbuktu. A bustling city, its marketplace is filled with the scent of spices and perfume, the chatter of many languages and clouds of dust stirred up by donkeys and camels. Traders examine wares, feel for quality, haggle over the cost. This is where the Dogon will get the best price for their indigo cloth as well as hippopotamus teeth, gold dust and ostrich feathers. But there's another smell in the air too: an underlying stench of fear from those who risk being sold into slavery and dispatched to a faraway land. The Dogon know the danger they face – one wrong step could end in enslavement. They soon hurry back to the shelter of their cliffs and the baobab trees.

But demand for cotton and indigo keeps growing, as does the threat to their way of life. The English, Spanish, Portuguese, Dutch, all seek workers to power their plantations, and Africa has been chosen to fuel their ambitions. Just the other day, a young girl was kidnapped as she gathered leaves for her indigo vat. Where is she now? Being transported overland to a port or already on a ship, chained up with other slaves, body upon body in a dark, stinking hull? One thing is certain: if she reaches her destination, she will be put to work and worked to her death. Now, the Dogon are starting to wish indigo wasn't such a mystery after all.

---

Indigo has long borne the weight of beauty and tragedy. From the plantations of El Salvador to the Dogon's dye vats, it symbolised the dual nature of enterprise. On the one hand, it was craft, connection to land, spiritual tradition; on the other, it was fuel for the twin engines of colonialism and commerce. As we trace indigo's journey across continents, it is a reminder of how an unswerving pursuit of material wealth can consume the lands and hands that create it.

The Dogon were forced to seek refuge in caves and baobab trees right up to the end of the nineteenth century in order to protect their way of life against the pull of empire, but today efforts like the Indigo Dogon project are beginning to rewrite this legacy. 'A strong pattern emerges when each knot is tied tightly enough to resist the dye's reach,' teaches master craftsman Mamoudou Nango in his indigo workshops. The same principle guides the project, which he leads. By intertwining local families with the plants that surround them – Malian cotton and indigo – they are creating a resilient community, one that thrives on local resources and knowledge, rather than on the exploitation of people and land.

Yet, while the Dogon are finding new paths, elsewhere the production of fibres and dyes continues to strip cultures of their autonomy, forcing entire communities to become cogs in the global economic machine. Marginalised and displaced people everywhere continue to fight for space, dignity and visibility. This brings me back to the question of how much value we place on textiles and their associated knowledge and skills, and at what cost are our clothes being made today? Just as indigo spread through the blue veins of trade routes and pumped millions of bloodied guineas into the coffers of empires, so we too might ask what invisible threads connect our own lives to contemporary stories of extraction. And how might we start to unravel them?

# PART FOUR
# Everything Is Connected

## Balance is not a place to reach but a rhythm to feel

*Ireland • 1700*

When we talk about balance, we often think of a definite state: in balance or out of it, something to attain or to lose. But the rhythm of nature isn't bound by such notions. Balance is not static; it flows freely, embracing stability and change. Equilibrium and flux are not opposites but two parts of the whole, constantly shaping the world that surrounds us, always in motion like the waves beneath my caravan on Branscombe's pebbled shore. The sea is a reflection of becoming, not being, the very essence of Abá philosophy. This is what makes nature so vibrant. And resilient.

Throughout history, we have seen how Indigenous communities understood this fluid balance, viewing plants not just as ingredients, but as members of the community, as integral as any human presence and full participants in the rituals of life. Mother Earth was not passive scenery, but a living being capable of agency, an equal partner in the act of creation. In this holistic worldview, every colour is a conversation, every fibre a strand in the story. Yet as the eighteenth century dawned, the world underwent a dramatic reshaping of understanding. Nature was reimagined. No longer a partner, it became a possession, inert and expendable.

---

It has been more than ten years since Isaac Newton published his *Principia*; ten years since we started seeing the world in a newly enlightened way. It turns out that the garden where God placed humanity is no snake-filled wilderness after all, but an orderly, civilised place, subject to rational laws. And if nature can be understood, it can be dominated, manipulated, exploited to maximum effect. No longer are we at the mercy of wild nature's whims. With new knowledge of how the world works, the barbarian wilderness can be harnessed to fuel human progress. Five hundred years after the word that means birth was born, nature is transformed into a shiny machine. And expediency replaces enchantment in the lexicon of progress.

As Europe's hunger for knowledge and power grows, so too does the tide of refugees, escaping old hatreds and bringing new possibilities. Among them are the Huguenots, fleeing French persecution of those of Protestant faith. The English fling open Ireland's doors for these strangers; not only will they help swell Protestant ranks, but they could teach those philistine farmers a thing or two about flax. Two years earlier, England abolished duties on Irish linen imports, part of a strategy to ensure a steady supply of the fabric, while slapping prohibitive duties on Irish wool to protect their own manufacturers. France's loss will be Ireland's gain as grand plans for mechanising the industry unfold.

Louis Crommelin, one of Ireland's newly arrived Huguenots, hails from a long line of French landowners with experience of growing flax. When the Crown learns of a man with knowledge grounded in experience, Crommelin is tasked with reporting back to King William III on the state of Ireland's linen industry.

He begins by travelling north to the elegant town of Lisburn in County Antrim, where trade is flourishing under the influence of local Quakers. At first glance, he imagines the task shouldn't prove too onerous, but it doesn't take long to uncover serious issues. Despite the region's fertile soils and long flax-growing history, the locals seem clueless when it comes to producing it properly. *Entirely ignorant of the mysteries relating to its manufacture*, he writes.[1] The women don't even know how to choose the right seeds or prepare the soil, resulting in fibres that are too short for high-quality yarn and, as for the habit of drying flax by the fire, he doubts even the best artist in nature could bring good colour to such cloth. To top it off, there are no standard measurements for reels of yarn, leaving honest workers at the mercy of those out to cheat them. Why do these problems persist? Crommelin anticipates the King's question and doesn't hold back. The real issue, he explains, is the deep prejudices that prevail. Spinning, weaving and bleaching are abject trades, lowly and poorly paid the world over. Anyone with skills or ambition avoids them entirely. How can people carry out such work without prospect of sufficient reward, and without decent pay, how will the industry improve?

The King is ruffled, although perhaps not surprised. He grants Louis Crommelin a royal patent and appoints him official overseer of royal linen manufacture in Ireland. Louis has big plans. He invites skilled Dutch farmers to teach flax cultivation, commissions spinning wheels from France, looms from Flanders and establishes standardised processes. His reforms promise increased productivity, higher earnings, although not everyone is happy with the rigid new system. This is, after all, an industry deep-dyed in tradition.

Next on Louis' list are plans to set up a bleaching green on the outskirts of town, a model of efficiency ready for replication.

How was he to know the memories bound up in cloth? For hundreds of years, women have made their way to the river, drawn like damselflies to its dawn-lit waters. Along the banks, they have laid down their linen and left it to develop a patina of dew. Everyone has their own secret bleaching recipe handed down through the family, using what is cheap and ready to hand. Fermented bran, buttermilk, wood ash, urine, even cow dung is employed in whitening cloth. Sun and dew, snow and rain do the rest. On frosty nights, some even go as far as bleaching with moonbeams, whitewashing their doors to harness every soft ray of light. Everything else has been taxed to high heaven, but at least the moon still runs free.

These women already have their methods, tied to community and land, and they're far from the mechanised world Louis Crommelin envisions. Progress is not as simple as a new bleaching green, a French spinning wheel or a newfangled loom. Louis is right when he talks about prejudice, but it runs so much deeper than wages. Their religion is banned, their liberties stripped away, under the everyday injustice of imperial rule. And now they're to become cogs in the Protestant machine? It's not that they don't need the work, and more money would be welcome, but they'd much rather take their linen down to the river for bleaching. This is their community. This is their church. This is where they sing and heal their hurts. If only the waters would run fast enough to wash the memories away.

---

I reach the end of writing this story and pause to prepare my lunch. But first, a walk to the foot of the cliff where the watercress grows in rivulets of water trickling onto the pebbles. It's a warm autumn day, so I sit outside on the caravan deck to eat my salad, bare-legged, short-sleeved, looking out to sea. Suddenly, at the edge of my field of vision, I glimpse something hovering, just above the tangle of madder growing up through my hedge. It's one of the peregrines, rarely this close. The bird is almost motionless in the air. And yet she's not, because I can see how hard she's working to keep herself poised at height with rapid shallow wingbeats, swooping a tight circle from time to time when a gust knocks her off course. In her constant dance between the forces of wind and gravity, the peregrine is the living embodiment of balance.

As I sit watching her, I'm reminded that survival doesn't always mean moving forwards; sometimes, it's about finding the strength to stay still. Today, with the planet tipping past the 1.5-degree limit of safety, I wonder if perhaps she is showing us the lesson we need most – that we too need to find a way to hold steady? As the peregrine surely knows, and we need to learn, the secret is not in conquering nature but in learning and practising, again and again, to live within its ebb and flow. Balance is not a place to reach, but a rhythm to feel. Becoming, not being. Like the women of Ireland who chose rivers over the bleaching green, we too can draw strength from stillness, the kind that is alive as running waters, as motionless as the peregrine. Because in pausing, there is power.

## How do you measure the earth?

***USA • 1750***

As Ireland threads between the future and past, and Louis Crommelin works to introduce standard measurements for reels of yarn, waves of Irish set sail for distant shores in search of a better life. There in the misty mountains where the four winds blow, straddling North Carolina and Tennessee, it is the land itself being measured by English hands. But not in any language its people know.

---

How do you measure the earth? By the fertility of its soil, the diversity of its vegetation, the height of its black walnut trees, the number of deer, the purity of the streams. That's how the Cherokee – those who call themselves the Aniyvwiya – measure the earth. These are not just physical measures, but reflections of *tohi*, the essence of life. The word means peace, and it means wellness as well, for the fast-running streams are their lifeblood, the rhythm of the earth, their heartbeat. To measure the earth this way is to understand its equilibrium, and to walk the right path is to maintain balance because when the earth thrives so does their community. Of course, they trade with the newcomers, exchanging deerskins, beeswax and natural dye cane baskets for knives, rifles, gunpowder, blankets, textiles and clothes. They adapt, they negotiate, but on their own terms.

Then men arrive on horseback, a covered wagon following on behind, its canvas made from hemp grown in the Virginia colony. They dismount and look around warily before unloading a mahogany box. Lined with flawless purple velvet, the soft padding encloses a glass-covered dial with a needle that spins lazily when they lift the lid, pointing to the four winds that blow across Cherokee land. The men have brought wooden stakes too, which they plunge into the earth, tying strips of coloured cloth to the tips. North, south, east, west, in every direction they pace, stopping here and there to kick aside a rock and dig another bright marker into Indian land, plotting straight lines through the quavering manna grass. These stakes are your boundary, the limitation of your territory, they say. On the other side of this line, the land belongs to the Royal Governor of South Carolina. If you dare disagree, we'll bind you to a bigger stake and burn you in the middle of your village. This is how we measure the earth.

How do you measure time? By the stirring of the black bears, the burrowing of woodchucks, the flashing of fireflies, the bugling of elk warming up for the rut. It's marked by the season for gathering black walnuts, green spheres that drop to the earth or hang heavily from the trees; if they leave them to rot, they'll be useless for dyeing. Soon the leaves of the walnut trees will start to yellow, a signal to the rest of the forest that it's time to shed their summer dress. This is how the Cherokee measure time.

But more men arrive, English, Scots, Irish. More wagons rumble in, women and children joining the throng. The settlers unload still more boxes, this time filled with axes. Now the Cherokee measure time by the rate at which the black walnut trees fall.

---

As in Ireland, English colonisation in America was accompanied by cultural violence: the Cherokee were forced to abandon their language, culture, dress and lands, culminating in their forced removal to Oklahoma on a journey remembered as the Trail of Tears. All to free up yet more territory for the settlers, as if what they had already taken wasn't enough.

Yet the earth remembers her old paths. The rivers keep flowing, the fireflies keep glowing, the black walnuts that survived continue to fruit. Perhaps balance is not lost, only resting, like the walnut, waiting for a signal to seed the future. To take root and rise, branching into a renewal of language, culture, rights and land. To reclaim what was stolen, rejoining people with the land of their roots.

## Eliza

*USA • 1759*

While the forests of the Cherokee are mapped out with stakes, the swamps of South Carolina have already been transformed into fields of profit. Into this changing landscape comes a young woman with a grand vision and persistence to match. Eliza plants more than seeds: she sows an empire of blue.

---

She was a scrap of a girl, but she made up for her size by the sweep of her dreams. My most valuable fortune and future happiness lie in my head, she always said.[2] And it turned out to be true. Plants were Eliza's first love, but when she was sent home to England to complete her education, the sacred kapok trees and blousy bougainvillea of her Caribbean home were nowhere to be found in her schoolbooks. While she understood an education was more valuable than any fortune she might inherit, she couldn't wait to return home to Cabbage Tree, the family's plantation in Antigua.

Colonel Lucas, her father, had different ideas. A sugar baron of wavering fortunes, he had set his sights on the South Carolina coastline, where – now that the original inhabitants had been driven out – the forbidding forest was being cleared from the coast to the mountains. The forest canopy and dense understory had kept the swampland beneath in perpetual shadow. But now black walnut trees, cedars and oaks capsized like a fleet of ships demasted in a storm of greed. Once the slaves had split, cut and piled up the wood, they dug out the roots that had anchored these ancient trees to the earth. 'Now that the vegetation has gone and the sun can reach these stagnant waters, this land will be perfect for rice,' Eliza's father enthused, skirting the matter of the substantial mortgage he'd taken out to purchase three plantations and a score of slaves. But when peace negotiations broke down between England and Spain, Colonel Lucas was forced to return to Antigua, leaving sixteen-year-old Eliza to manage the half-tamed wilderness on her own.

The Lucas plantations stretch along a winding salt creek, named for the Wappoo Indians who once inhabited the coastal lowlands. Where they have gone, no one can say. Rice is the obvious crop to grow in these swamplands, just as her father suggested, but Eliza has a restless mind and playing it safe rarely yields the greatest rewards. She pores over her father's botany books, absorbing their wisdom, knowing this is an opportunity to put her obsession with the vegetable world to good use. Then she starts experimenting, trialling different plants to find those best suited to the humid climate and quaggy soil. Ginger, cotton, alfalfa and cassava all grow to perfection under her watchful eye, but it is indigo for which she harbours the greatest hope. When the first year's crop withers on the stalk and the plants' grey heads bend to the frost-bitten earth, she sheds tears of frustration but doesn't give up. The next harvest fares little better, but failure isn't an option; the plantation's debts loom large. She writes to her father requesting more seed.

Neighbours eye her efforts with scepticism. *Tell the little visionary come to town and partake of some of the amusements suitable to her time of life*, they laugh.[3] Others grumble she's just wasting good land, for not even experienced farmers can coax the crop through a Carolina winter. But Eliza is stubborn as a black walnut root and, in any case, she loathes the endless round of tea dances in Charleston. For six years she sows indigo and, while she waits for a successful harvest, she quietly teaches her slave girls to read, training them to pass on this knowledge to other children. (Future plantation owners will take a less enlightened approach because slaves that read abolitionist pamphlets may plan a revolt. Seditious literacy will be banned throughout the state.)

At last, Eliza's persistence is rewarded: twenty weight of indigo is harvested, with ten more on the way. Production of the precious pigment can commence. Of course, she hasn't spent all this time waiting idly for a good harvest. Three years earlier she sent to Montserrat for an overseer who was familiar with the process of transforming plants into pigment, but the blue lumps he produced were of such poor quality, Eliza began to suspect subterfuge to prevent her indigo rivalling that from the West Indies. You're adding too much lime, she pronounced, sending him packing. Determined to master the art herself, she studied every account of indigo production she could find, comparing methods, adjusting timings and, as she experimented, a pattern emerged: what once seemed like alchemy was little more than a recipe.

Now she gathers her slaves and explains all the steps: the steeping, the fermenting, the constant watching until bubbles rise as if from an underground spring. She shows them how to siphon and strain, moving the mixture from one vat to another; these vats alone have set her back the cost of a new adult slave, not to mention the expense of constructing irrigation channels and drying sheds. Worse, the stench of rotting plants rises in thick waves, drawing flies by the thousand and making her men sick. They have no choice but to endure the conditions: the flies, the sickness, the heat day by day. But Eliza will not stop now; she holds her breath for a week. She needn't have worried about the plants' potential: indigo is a shared language across the world and her slaves recall its bloom from their own villages. Before long her blue pigment is on a boat bound for England. Fortune favours the brave, so they say, and indigo doesn't disappoint Eliza, although her workers may well have wished for a different harvest. The ongoing war with Spain has severed the usual supply routes and indigo is a scarce commodity, meaning her small blue cakes command premium prices.

Eliza is not just a bold thinker, she's a pragmatist too. Rice, she explains to her neighbours, is all well and good as a staple, but it only provides income for half the year. The rest of the time our labourers are idle. England needs dye as its textile mills multiply, and it goes without saying they'd rather back their own settlers than the French, so why don't the other farmers build on her blueprint? She begins sharing her seed and, as South Carolina's indigo business flourishes, so does demand for labour, until newly arrived slaves command a higher price than anywhere else in the colonies.[4] These men and women bring not just sea-weary bodies, but minds churning with ancestral knowledge of planting, soil and the mysteries of the indigo vat. Many of the new arrivals have passed through the hands of African slave traders in exchange for a length of indigo-striped cloth.

Indigo had enriched their community, but Eliza is practical enough to see the need for diversification, turning her hand to flax and hemp, bringing in indentured Irish servants, a spinning wheel and loom. Soon mulberry trees line her fields, silkworms feasting on their leaves, reviving an earlier idea for the colony that had fallen out of favour. And all the while the indigo trade keeps growing, becoming one of the colony's major exports, second only to rice. As Eliza's ventures expand, so does her family; she marries Charles Pinckney, a respected lawyer and politician, and four children arrive in rapid succession. But they are not the only fruits of her labour: fig orchards flourish beside oranges, nectarines, peaches and plums. Stands of oak and cedar rise where barren land lay. Eliza mourns the frenzied clearing of the swamps, seeing in a way few others do the balance nature requires. In her agricultural revolution, all plants play their part.

Sometimes, when the mulberry trees are in bud and the mockingbird in full song, she sits back in her rocking chair on the porch and allows herself a moment of reflection, marvelling at the dreams one can sow from a few scorned seeds and an inexhaustible love of plants.

---

Eliza Lucas Pinckney's botanical research and unflagging spirit are a lesson in enterprise. For a woman not only to build a successful business but to transform the economy of an entire region would be remarkable at any time, even today. Yet an empire is never built by one person. In the shadow of Eliza's success, retold through letters to her family and compiled by her niece, lie the unheard stories of hundreds of hidden labourers who turned her vision into reality. Indigo was rooted in the toil of the enslaved and watered by the sweat of bodies traded like commodities, each seed that sprouted pressing the weight of slavery more deeply into the South Carolina soil. They are silent partners in Eliza's success. And this is their story as much as hers.

## The end of the forest

***Paraguay • 1773***

Deep in the jungles of Paraguay, the Jesuit missionaries were faced with an impossible choice: cling to their pacifist creed or arm their followers and fight? It was a decision that would cost them everything.

---

At first, the Jesuits, the Companions of Jesus, were bewildered by those who called themselves Abá, The People. In vain, they searched for temples, altars, idols, for something to tear down, convinced that by uprooting the old ways they would prepare fertile ground for the seeds of Catholicism to flourish. But they found no sacred structures: no walls to destroy, no totems to break. The missionaries even struggled to uncover a name for their God, a bridge of words to communicate the divine, eventually settling on Tupã, god of thunder, guardian of the paradise into which this Earth was born. Conversion remained the Jesuits ultimate aim – as many souls as possible for Christ – but gradually a new vision took hold of a bold social experiment, the creation of a utopia no longer attainable in the Old World. Their missions would be more than a refuge; they would be an audacious reimagining of society itself, with faith as their foundation, equality their creed and self-sufficiency their lifeblood.

Blinking, the Abá emerged from the womb of the forest, forced out into the bright mission gaze. Those who accepted rebirth through baptism were known as the Guaraní, warriors in name and spirit. But civilisation, the Jesuits insisted, required more than faith. They must be schooled in the arts and learn useful skills such as agriculture, leather tanning, carpentry, sculpture, printing, textiles, musical instrument-making and more. Beyond this, they decreed, the Guaraní must wear clothes, for this is what separates us from nature. Haven't men and women covered themselves since Adam and Eve departed the Garden? Look at the unconverted who still wander the forest naked, untamed, always on the move. What the Jesuits initially failed to grasp was that the Abá saw no divide between themselves and the emerald-crowned forest, no separation between the sacred and the profane: they were as two strands plied in the same endless thread, flowing as one. Their paradise – Ywy Marã Ey, the Land Without Evil – was a place to be found here on this Earth.

To fulfil their vision of salvation, the Jesuits set the Guaraní to work, cultivating cotton to be made into cloth, not just for their own use but for export to Europe, because the business of saving souls doesn't come cheap. But as the Jesuits wove their designs, slave-raiders from Brazil arrived to unravel their efforts. Cotton had become the lifeblood of Brazil's economy, so essential to trade that in the northern regions, municipal councils even set prices in rolls of cloth. As European demand continued to grow, more hands were needed to sow and harvest the crop. These mission lambs would be easy prey, so the slave traders thought, drawn from their homes in the forest and conveniently corralled within Jesuit settlements, vulnerable and exposed. But the slavers hadn't counted on the missionaries' resolve to protect their flock and conserve their forest sanctuaries, whatever the cost.

First, the Jesuits and their congregation retreated from the coast, a mass exodus of ten thousand or more, moving southwards, inland, towards the distant mountains where the Abá had once searched for the Land Without Evil. When flight no longer protected them, the missionaries made a hard choice. It was time to set aside their philosophy of peace for a more militant form of paternalism. Laying down their artisanal trades, they turned to making gunpowder instead, sending to Buenos Aires for guns and for soldiers to train their flock in the tactics of war. The Jesuits were accomplished at most arts to which they turned their hands and commanding field armies proved no exception. The Guaraní lived up to their new name: the warriors repelled raid after raid with a ferocity equal to the slavers' greed. The Jesuits made for a powerful enemy, but their armed resistance, their monopoly on trade and their vision of a self-sustaining paradise had ruffled the feathers of colonial and religious authorities alike. Such utopian ideas were a dangerous provocation. After more than a century, the Crown forced the Jesuits to abandon their transformative mission and leave South America.

Now, the churches and workshops lie in ruins, abandoned to the tangled vines of the jungle – a haunting reminder of a broken utopia. The forest has swiftly reclaimed the land, but here and there you can still find traces of a dream the world failed to understand. After the Jesuits returned to Spain, the Guaraní dissolve into the forest, returning to their former haunts, searching for the old paths their ancestors trod. Drawn by lingering memories, they resume their search for the Land Without Evil, guided by reciprocity and the unfolding power of the word. Words are the foundation of their existence, not written, but alive. Their very essence is woven into the word-soul, because for the Guaraní, speech is the fabric of being, every word a footstep closer to unravelling their destiny as they press deeper into the forest.

Always they are moving, faster now, carrying memories through shadows and light, forever tracking the *uru'ku*, the gleaming red seeds that illuminate their path. But these bushes are vanishing, as are the trees. Where once flowed a realm of unbroken green, the land lies splintered, scorched, stripped of life. The levelling converges from both east and west as cotton and cows supplant trees. This, they know, is not simply destruction; it is *mba'e maquá*, the ultimate evil. For when the forest dies, so does the future.

---

The Jesuits took a fateful risk in siding with the Guaraní, yet they knew that benevolent neutrality wasn't an option. Some may say it wasn't a risk worth taking, that it cost the Jesuits everything. But can we be so certain? What if they had done nothing at all? Choices made centuries ago shape the struggles of today. Displaced from their ancestral lands by monoculture crops and facing high suicide rates among the young, the Guaraní's search for Ywy Marã Ey, the Land Without Evil, has taken on a tragic poignancy. Yet still they resist, adorning their bodies with uru'ku, weaving fine cotton cloth and spearheading grassroots initiatives to reclaim what is theirs. The Land Without Evil remains an empty space they still need to fill.

For the Guaraní, as with most Indigenous Peoples, land is everything. As I read about ranchers seizing their territory, burning homes and driving communities out at gunpoint, about families forced into makeshift camps beside the highway or corralled into tiny reservations, about agribusiness giants linked to human rights violations, I wonder what risks we are willing to take for justice, for our vision of utopia, even when it stands at odds with the prevailing worldview. Today, these questions sound louder than ever. What sacrifices will we make to speak truth to power? Are we willing to disrupt entrenched systems, even rethink our ideologies? We stand, as the Guaraní did, at the edge of a dying forest. We too have a choice: to retreat or resist.

## C'est puce!

*France • 1775*

At its root, fashion is a universal language. A way to create, shape and express identity, values and culture, or as the French say, *façonner*. More than the clothes that we wear, it reflects how we weave our identity into meaningful forms, deep-dye our values with purpose and imbue our traditions with long-lasting hues. As the Guaraní moved deeper into the forest, trusting the red *uru'ku* seeds to guide them to their future, across the ocean, France's bourgeoisie were also looking eagerly to what lay ahead, although for them the future shimmered in the glint of gold mirrors and folds of silk gowns. Versailles was abuzz with fashion's newest obsession.

—

Every season has an It colour, and for summer '75 it's puce. An earthy, pink hue, some say it resembles the 'Cuisse de Nymphe Émue' roses that grow in the palace gardens, but with the softness of age, the very look Rose Bertin aimed for when she designed Marie Antoinette's latest gown. Each year, it's a challenge to rustle up something new, for whatever the Queen wears will ripple through France like a silk ribbon. Not that there isn't some grumbling at court when her taste veers too far towards novelty. This time, Rose knows she has nothing to fear: puce has hopped from Queen to Court with the easy grace of the insect that inspired its name. Soon, all Paris will be draped in the colour – there's a reason she is known as the Minister of Fashion. Although this time, she can't take all the credit.

'What do you think of Rose's latest creation?' asked the Queen, twirling in front of her husband. Her dress flowed around her form, shimmering as it caught the light.

The King rolled his eyes. 'C'est puce! It looks like flea droppings on the bedsheets.'

For a moment Rose held her breath. Even the courtiers dropped to a hush. But the Queen simply laughed and spun faster, her gown billowing like clouds about to drop rain. 'I'm wearing flea colour!' she sang, and Rose breathed again.

It was Madame de Pompadour, the mistress of the King's grandfather, who popularised the pink hue that came to be known as Rose Pompadour and, when Marie Antoinette arrived at Versailles, she too was enchanted by the shade. Now, the Queen's private garden boasts two thousand dog roses as well as cabbage roses and innumerable other varieties, all filling the air with their warm heady scent and many blooming in the celebrated pink. Inside the palace, rose garlands criss-cross the rooms; even Marie Antoinette's bedroom is papered in the sensual shade. Rose Bertin has timed her introduction to perfection, coinciding with the

Queen's twentieth birthday. Celestial blue was all well and good for a virginal princess, but roses have forever stood for beauty and love. Pink is a colour befitting a Queen. Not that her gown is strictly pink, of course, but neither is it mauve, nor brown, nor grey. Puce is a colour in between, delicate and strange. It is an oddly apt name for this hue, Rose decides, and will be all the more fashionable for being christened by the King.

No one can take their eyes off the Queen as she walks the palace halls in her modish new gown, and puce advances through the courtiers like an infestation of fleas. The following morning, Versailles is emptied of valets as carriage after carriage rumbles down the track, dispatched to Paris to secure velvet, silk, sateen, any cloth they can find in that blood-sucking shade because all women must wear puce for the next grand ball. Even men fall in love with the colour. Buy a puce jacket, it's rumoured, and you'll fast become the Queen's favourite. By dusk, the price of puce has doubled.

Puce keeps more than the dyers busy. In Provence, fifty water mills turn night and day to grind madder roots into powder, while merchants scour France and Italy to replenish dwindling supplies. Metallic salts are needed too, not only to help the dye affix to the cloth, but to achieve the distinctive hue. Meanwhile, some work with New World dyes, making puce from pearl ash and annatto – derived from the very same seeds the Guaraní call uru'ku. Others give the cloth a good soaking in alum, bathe it in brazilwood and finish it with logwood from Mexico. Still others reach for the green orbs of the black walnut, the same fruit harvested by the Cherokee before their trees were chopped down.

Botanical colours are fickle, changing with the container, the mordant, the soil, the seasons, even the phases of the moon; any number of factors can change their tint, not least the haste of dyers struggling to keep up with demand. In a matter of weeks, one shade becomes many: old flea and young flea, live flea and dead flea, flea belly, flea thigh, flea head, flea back… every tint displaying its own subtle shift, each contributing to the fever of consumption. Flea swiftly makes the leap to the bourgeoisie, who appreciate its ability to disguise dirt and stains. Puce takes Paris by storm.

Rose Bertin smiles to herself, knowing her star has not dimmed, and sets to work dreaming up the next must-have trend.

---

As the forests of the Guaraní were being erased by axe and flame, the gilded halls of Versailles rustled with annatto-dyed gowns. Derived from the red seeds of uru'ku, the shrub we commonly call achiote, the Guaraní's sacred beacons were transformed into fashion currency for the elite. Puce's moment of opulence may have been fleeting, yet its rainforest roots speak to the kaleidoscopic ecosystem that textiles navigate, flowing through time like a river, stories layering upon stories like sediment. Sometimes they settle in unexpected places, far from their origins.

Two years after my research voyage to Easter Island, I found myself once again back on the water, but this time I was far from the vast blue Pacific. Our boat was a tiny vessel, flat-bottomed, barely fifteen feet from bow to stern, and we were surrounded by woodland and rolling hills. We were on Rudyard Lake, a two-minute stroll from my front gate. Although the setting couldn't be more different, my mission remained the same – to uncover the mysteries below the surface and work out where fashion and textiles fit in.

The day was bright and cold. We'd been out on the water all morning and were ready for lunch, but the anticipation of what we might find kept us going. During the Industrial Revolution and beyond, textile manufacturers flocked to the nearby town of Leek due to the River Churnet's unusually pure water. Soft to light, dyers called it, perfect for gently infusing colour onto cloth. But as dyeworks began pumping effluent into the river, it gained a murky reputation as Europe's most polluted waterway. Mills sprang up too, and still dominate the skyscape today. Big Mill once employed two and a half thousand people, but the machines rattled out their last bolt of cloth more than fifty years ago. Upstream of Rudyard Lake, textile manufacturing was poorly documented, but we knew there had been mills, both cotton and paper, and a colour mill said to have produced black dye from coal dust. Could this perhaps have been used for Leek's famous Raven Black? Favoured by Queen Victoria, the colour shimmered on silk like a raven's wings.

In the boat with me were three scientists, Deirdre, Toni and Tom, all keen to explore how the historic production of textiles affected local water ecosystems today. 'But wait!' I hear you say. 'Don't natural fibres disappear into nothing within months, if not weeks?' Well, that's what I used to believe too. Instead of the clamshell device I had used to scrape the seabed in the Galápagos Islands, we lowered a clear cylindrical core sampler, designed to penetrate the layers of sediment. As the tube was hauled back to the surface, the anticipation was palpable. What secrets were hidden in the depths of the lake? As we turned our boat towards shore, none of us realised how important these twenty centimetres of lake mud would be.

We had one last visit to make. The Reverend Geoffrey Staton, who had spent years working in Leek's dyeworks, was waiting at my dear friend Judy's house to tell us about his experiences. Geoffrey settled in a chair and began to recount his days as a dyer's apprentice in 1955. There is a saying that you can predict the It colour for the season by looking at the rivers in China. Well, it turns out the same once held true for Leek. Geoffrey described how the Churnet changed hue two or three times a week depending on the dyes pumped out by the sprawling dyehouses lining its banks. Stories even circulated about catching purple fish. Did remnants of this industrial pollution still linger? This is what we hoped to find out.

On a bright April morning a few weeks later, a group of eager volunteers gathered at Rudyard Lake visitor centre. Memories of childhood frogspawn hunts resurfaced as they arrived with jars and glass bottles filled with water from different spots along the river. Our sediment samples were there too, extruded and sectioned into Petri dishes, one for each centimetre of lake deposit. 'Sediment at the bottom of lakes acts like a time machine, revealing changes in fibres and pollutants over the years,' explained Tom, as sunlight reflected rippling shadows onto the walls. We sat at our microscopes and began searching for fibres. I rotated the focus knob, peering through opaque sludge, brown as over-steeped tea, until a wavering line caught my eye: a substantial blue fibre with a clear twist, most likely from a pair of jeans. Beside it stretched a smooth, hair-like strand that appeared brown, but could have been any colour before being mired in the mud of years. That one looked synthetic. And the fuzzy yellow piece must be wool. As the day drew to a close, our tally of fibres had far exceeded our expectations, and I wondered what these stale remnants of cloth could tell us, suspended in the dregs of the past. The samples were packed off to Northumbria University's forensic science team for further analysis.

Months later, the results came back. They revealed a complex story. Synthetics were indeed present, but very few. The core sample was mostly comprised of cotton, a seemingly natural material, with a small amount of wool. As the research team pored over the data, comparing lead dating to radiocarbon dates, juggling countless calculations, there seemed little doubt what the results were telling us: some of those fibres had been lying in Rudyard

Lake since the final years of the Industrial Revolution. My mind reeled at this revelation, and it raised a new question: could our well-meaning efforts to switch to natural fibres be unknowingly causing a different environmental threat? And if so, why hadn't we known this before? It turned out we simply weren't looking for them, dissolving everything natural – organic matter as well as natural fibres – to reveal the plastic. As I thought back to the microscopic world of sediment samples, I remembered those coloured lines snaking through the mud in my Petri dish. They looked harmless enough, but those seemingly benign fibres could be toxic time bombs, the effect of dyes and other chemical treatments lingering for years. If dyes could impact human health and living organisms, could they be affecting biodegradability too? Would that blue strand have survived the years if it had been dyed with natural indigo instead?

With countless unanswered questions swirling around my mind, I resolved to find a new way of working, one that acknowledged the interconnection between different strands of life. A new colour narrative where fibres and dyes were not the root of our problems, but catalysts to help grow solutions.

## A world entwined

***Venezuela • 1800***

Last night was a Hunter's Moon, a huge pearl in the sky, casting a pathway from the door of my caravan to the distant horizon. So bright was it, I woke up in the middle of the night thinking I'd left a light on. When I looked outside, the sea was awash with innumerable sparkling pinpoints, as if lit by diamonds. Unable to fall back asleep, I opened the window to welcome the sound of the waves and a ribbon of moon. After tossing and turning a few minutes more, I succumbed to the inevitable: I scrolled.

The first post on my Instagram feed was from Amazon Frontlines, showing Ecuador's shrinking Carihuayrazo glacier. Formed over twelve millennia ago, around the time the earliest hunter gatherers arrived in modern-day Ecuador, the glacier has lost up to ninety-nine per cent of its surface in the last seventy years. The name of the slumbering volcano on which the glacier sits comes from the Quichua language. I knew *cari* meant man: a homonym of my name, it's a source of amusement among my Ecuadorian friends. In the morning, I messaged my friend Willian, with whom I chat in my rusty Quichua from time to time. He explained that *huay* is the word for wind, and *razo* means snow or ice: the man of icy wind. But the glacier that mantles this volcano has almost vanished. Why? Climate change, of course. But it's also being caused by smoke drifting from the Amazon as it burns. Carbon deposits reduce reflectivity, accelerating the melting, increasing vulnerability. Everything is connected.

---

*The first colonists very imprudently destroyed the forests. Evaporation is enormous on a stony soil surrounded with rocks, that radiate heat on every side*, observes Alexander Von Humboldt, Prussian baron, naturalist and explorer as he crosses a low-lying mountain chain after departing Caracas.[5] Along with his travel companion, botanist Aimé Bonpland, Humboldt intends to spend five years exploring these lands of mighty flora and great civilisations, and Venezuela is where they are to begin their journey across Latin America.

Humboldt tries to channel his thoughts into the measured lines of a travel journal, but it's of little use: the book reflects how his mind works. There are sketches, detailed calculations, marginalia (if he remembers to leave a margin that day) and plenty of crossings-out. He starts writing in German, then switches to

French, adding comments in Portuguese, Spanish, Italian, Latin, Greek, English. Sometimes he adds words from Indigenous languages too. His sentences are tangled, cramped, like the profusion of thoughts filling his mind. He could, of course, express himself perfectly well in German, but language is more than a jumble of grammar and sound; its inner form represents a way of seeing the world. Like the Guaraní word-souls, language binds nature to ideas.

Three months into the new century, Humboldt and Bonpland are standing on the shores of Lake Valencia, Venezuela's largest freshwater lake. The rainy season has already started, yet the lake's level remains low. The settlers are worried. Not long ago, their colony's indigo production rivalled that of lands to the north, but now one of their most lucrative crops will no longer grow without artificial irrigation. For that, running water is needed and this presents a problem because the springs have dried up, as have the river beds, and still the level of the lake falls. Until it rains – and how it can rain in Venezuela! Then rivers turn to torrents, furrowing into the hillsides, shaking loose the topsoil that no longer has tree roots to hold it fast, ravaging the countryside.

This, Humboldt knows, shows the interconnectedness of nature. *In this great chain of causes and effects, no single fact can be considered in isolation.*[6] For thousands of years, lakes have achieved an equilibrium between the amount of water flowing in and that lost through evaporation and filtration. This disruption is likely due to recent local factors, he pronounces. Fifty years ago, the slopes of these mountains were covered with trees – mimosa, ceiba, fig – shading the earth from the withering effects of the sun and reducing evaporation. Now those trees have been felled, the heat is like midsummer in Naples, and water, once abundant, is fast disappearing. They are all inextricably entwined, he explains, the cutting down of the forests, the drying up of the waterways, the floods, the cultivation of cash crops. Especially indigo, which impoverishes the soil more than any other. *The lands of Maracay, Tapatapa, and Turmero, are looked upon as exhausted; and indeed the produce of indigo has been constantly decreasing*, Humboldt scribbles in his journal.[7] This calamity – the felling of trees, the depletion of soil, the exploitation of land – will not easily be undone. Indeed, its impact will be felt for generations to come.

Humboldt understands that nature is not a machine: it is a web of life, a self-sustaining network embracing everything from the smallest insect to the tallest tree. Or it would sustain itself if we would just let it be. Carry on like this, he warns, and the soils will resemble those in Spain, parched and bald, scalped of their green pelt. Alexander Von Humboldt sees connections everywhere. Everything hangs together. If one thread is pulled, the whole tapestry may unravel.

---

Humboldt saw it more than two centuries ago, the fragile balance, the intricate web. How one thread of nature, when pulled, interrupts the pattern, loosens the rest. How our desire for colourful clothing brought scarcity elsewhere *as the wants and restless activity of large communities of men gradually despoil the face of the Earth*.[8] He tried to warn us, but we carried on cutting, digging, burning. The Carihuayrazo glacier is disappearing for the same reasons Lake Valencia dried up. The same reason Çatalhöyük's landscape turned to desert eight thousand years ago. Human exploitation of nature was altering the environment in far-reaching ways, including the climate. Little has changed in the past two hundred years, except now the destruction is amplified; chainsaws and bulldozers make matchsticks of the forest, fuelled by unchecked consumption and driven by the insatiable engines of trade.

It begins with a single thread: one unnoticed snag, a delicate weakness exposed. The unravelling may be hidden at first, but soon a chain reaction is set in motion, compromising the integrity of the whole. What was strong and whole starts coming undone, and in our current climate the tensions pull harder. To watch this happen can make us feel helpless, witnesses to loss. The task of weaving it back together can seem impossible. Yet isn't this the essence of human tenacity? What we have done time and again over the centuries, millennia even: to gather the scattered threads, to reunite what has come apart, to create something new from the frayed remains of the old. Nature and humans working together in a tireless act of repair. Each interweaving is more than an attempt at preservation: it is an act of resistance against those who seem hell-bent on disarray.

In my own way, I realise that I, too, am trying to cultivate new connections, weaving something frayed back together again. After years hovering under fashion's spotlight, I have been drawn back like the tide to this place where I write, to solitude, to the Undercliff, where I'm learning the names of the wildflowers and fostering new relationships at the margins of land and sea – with the world around me and with myself.

Sometimes we weave together. Sometimes we weave all alone. Sometimes we weave with laughter or crying. Sometimes we weave in silence. The important thing is to weave a little more every day. Together we lace the threads of connectivity. Together we rewrite the book.

## F for flax, G for gun

*New Zealand • 1820*

Thousands of miles away, in the land the Dutch called New Zealand and the locals knew as Aotearoa, Ngāpuhi elders understood how one small action could disrupt the whole. With its sword-shaped leaves and waxy red flowers, *harakeke*, the plant Europeans named New Zealand flax, has long been woven through the fabric of their community. But harvest it carelessly, everything could start to unravel.

---

*Hutia te rito o te harakeke, kei hea te komako e ko?* If you pull out the centre shoot of the flax plant, where will the bellbird sing? Harekeke, the flax plant, grows from its core, fanning out from a tender young heart. As new leaves unfurl, they are shielded by parents, surrounded by grandparents, extending the circle of life. Take care of these plants, caution the elders, for they are a reflection of the community into which you are born. However great your need to harvest these leaves, if you pull from the middle, the plant will not regenerate; it will rot and die. Then where will the bellbird sit and sing? Do as we have shown you and cut the outer leaves near the base with a sharp mussel shell. Here in Aotearoa, they've no need of Humboldt to reveal the truth that everything is connected.

For centuries, the Ngāpuhi followed these teachings. With nothing more than mussel shells, they stripped fibres from the flax plant's tough outer leaves. Mostly these would remain in their natural shade, but colour imbued their weavings with meaning: when dipped in iron-rich mud, the fibres turned black, in *kōkōwai* clay they bloomed in pinks and oranges, *tanekaha* brewed earthy browns, while with *kanono* bark they were golden. Then, with little more than their fingers and a pair of upright sticks, the weavers created their ceremonial cloaks; although don't picture any rustic loom – it took years to master this intricate art form. Borders were finished in the *tāniko* technique; twisting weft threads around the warp, they interlaced not just pattern, but identity and values. Lastly came the adornment: kiwi feathers dappling the folds, or thousands of flax tassels rustling and swaying like foam on the sea. Flax made skirts and rain capes as well, adorned with leaves and vines for texture and contrast. But harakeke offered more than beauty; it served every purpose under the sun. Eel traps, baskets and snares fed communities, and the fishing nets were so wide they took a hundred men to haul in. Harekeke was life sustaining.

How times have changed. The swing, when it came, was violent. The cords that bound people to plants, community to land, have twisted and turned into something unrecognisable, entangling and ensnaring in entirely new ways. This is not because young people have stopped harvesting flax – far from it! In fact, they can't get enough of it. But it's not used to adorn bodies or make traps and nets to fill bellies. None of these are made anymore; not since the Ngāpuhi began bartering their precious fibres for white man's goods. Flax has become currency, and the fibre that symbolised life is traded for death. One ton of scraped flax buys a musket; three tons secures an old Brown Bess. Villages hum with the rhythm of flax being stripped, fibres being beaten, all to put another hand to the trigger in the fierce tribal wars. Some have even abandoned their homes, moving closer to the swamps where the precious plant grows. For if it wasn't for the flax, the Ngāpuhi would have no means to buy guns. And without guns, they wouldn't be able to spread fear and panic among neighbouring tribes, driving out their enemies, expanding their territory, because anyone living between their lands and the sea might stop them spinning a profit. What do the white men want with all this flax, some wonder, as they try to imagine a land where harakeke doesn't grow. How many flaxen cloaks do their rulers need? Little do they know that most of their fine fibres will end up as cheap rope.

Flax has become more than a fibre for the Ngāpuhi. It is a lifeline, vital for their survival in this shifting world order. Flax protects their interests, which align with those of the white men down on the coast: the missionaries, the sailors, the whalers, the slavers. Men with hungry eyes and a keen appetite for profit. Hongi Hika, the Ngāpuhi chief, even sailed the oceans to meet King George of England himself. In this meeting of worlds, Hongi presented the King with a red *korowai* cloak, a piece of his homeland woven from flax. In return, he received symbols of European power: a gleaming double-barrelled flintlock musket, a suit of armour, a helmet and sword. A cloak of flax for the ceremony of war: that's how the white man's exchange system works. Armed with these weighty tokens, Hongi returned to his people, where each musket meant greater authority, and the means to capture slaves to grow ever more flax. Slaves too have become part of the economy: five strong men will buy a Brummagem musket, churned out in the Midlands town known for bottom-rung weaponry. For slaves who don't make the grade, there's a lucrative local trade in smoked heads. Their brains removed, steamed, dried and smoked over fire, these once sacred symbols of tribal negotiations have become collectable curios for Europeans, who exchange them, inevitably, for muskets. Flax and guns keep the Ngāpuhi alive and send their neighbours to the grave in a puff of white smoke.

In this frenzy for flax, the Ngāpuhi are blinded to the old wisdom and forget their elders' teachings. Forget that if you pull out the centre shoot the plant will die. No one is listening now; even if they were, the elders' wise words would be lost to the thunder of guns. And what of the bellbird? Does it still sing? Listen closely in the stillness before dawn, and you might just catch its song before the crack of another musket shatters the air.

---

When cravings for wealth and power eclipse the value of human dignity and the thread that binds humanity to nature becomes severed, the consequences are devastating. Two hundred years on from this story, the same challenges remain: how do we divert the sharp mussel shells of those who through deforestation, industrialisation or extraction seek to slice through the fabric of life? And how can we protect the bonds that unite community and earth? In New Zealand today, a new generation of Māori artists and designers are doing just that, picking up half-forgotten strands of harekeke weaving, along with their language and heritage. When they twine in the tāniko technique, they reinforce not just the borders of garments, but Māori rootedness in culture and place.

After his explorations in Latin America, Humboldt expressed the belief that a better understanding of nature could help us see the connections between people and places, present and past. This seems a good place to start. Just as New Zealand flax regenerates from the centre, perhaps our natural systems can regenerate if we commit to sustaining them, always nurturing the heart.

## New things for good purposes

***England • 1839***

While my father's side of the family have been stubbornly Branscombe since records began, my mother's branch is far more itinerant. One man who looked out at the horizon and saw the future was my third-great-grandfather, Charles Claude Hamilton. Born from the unlikely union of Madeleine Collet, an orphaned French Catholic lacemaker, and John Hamilton O'Hara, Protestant Squire of Crebilly Castle and vast Irish estates, Charles Claude's life was shaped by opposing forces from the start. After his father's death, the legal battle over their inheritance became a cause célèbre, with Daniel O'Connell – future liberator of Ireland – representing Madeleine and her sons, pitted against a powerful establishment intent on suppressing Catholic emancipation. All it needed was a few bribed witnesses to sway the fully Protestant jury. The court declared the marriage void, their sons illegitimate, and Charles Claude was cast adrift, a victim of the sectarian tensions that defined the age. For all its dramatic scope, the outcome left my family on the margins, their hopes undone.

Yet while the authorities tried to tether his lofty ambitions to the earth, my third-great-grandfather found a way to rise above, escaping the world of shadows and allowing his imagination to soar. Charles Claude envisioned a future where limitations could be overcome not by money or privilege, not by holding the keys to the castle, but through the power of creation and the sheer will to dream.

---

*It is an incontrovertible fact*, Charles Claude Hamilton writes from his London home, *that the nineteenth century is already, before half its current era has elapsed, as brilliant in important discoveries of science as any of the most famous periods, either of antiquity or modern times.*[9] His graphite pencil moves swiftly across the page, although his hand hesitates now and again, suspended in uncertainty, as if measuring the weight of what he wants to convey. The glow from the coals casts a shadow across the parchment, but his words are illuminated by more than the fire; they are outpourings of a radiant dream.

Charles Claude has, by this time, already picked up the cube of Indian rubber on his desk and erased the lines once, admitting to himself the absurdity of his proposal, hearing the cries of ridicule. After all he has been through, he is fearful of deep-rooted prejudices that give rise to disbelief. Who will listen to a man who once stood at the centre of a national scandal, his claim to noble birth dismissed, his hopes in tatters? Since being turfed out of his ancestral home, he can barely make ends meet without the Royal Literary Society's support, and there's always another child on the way. Yet this same man who has felt the barbs of bribed witnesses, heard the sneers of biased juries, cannot quell the urge to write it all out again. He will not be bound by the smallness of other people's imaginations. *New things, by permission of God, for good purposes*, he continues. *This is the true nature of invention.* The innovations of tomorrow are no idle curiosities but acts of bravery for the betterment of all humanity.

The possibility of flight, that most ancient of dreams, has long eluded the contrivances of great men. Charles Claude understands only too well the failure of Icarus, but to his mind the sole impediment to the art of flying has been in acquiring the proper materials. With the recent arrival of a wondrous new substance, this castle in the air may have found its building materials upon earth. Born of the steaming rainforest of Brazil, Indian rubber, he believes, will change the way we see the world.

Across the seas in the forests of South and Central America, Indigenous Peoples have tapped latex from trees for millennia, combining with juices from morning glory vine to make the material less brittle. Yes, they used it to keep themselves dry, but mostly they employed it for another high-flown purpose: to make rubber balls for their favourite game. Then, just this year, the captain of a windjammer sailed from Brazil into Boston harbour with a pair of waterproof rubber boots. When the cargo was brought ashore, an order for five hundred pairs swiftly followed. Now waterproof fabrics are being made in New England's textile houses. Even in England, Mr Macintosh has started selling his patented rubberised coat, and Mr Nadier has woven rubber threads into cloth, stating his intention to make clothes and accessories.

It is this same Indian rubber or *caoutchouc* that will clad Charles Claude's vision, the missing ingredient to allow humans to take flight. He sees it clearly this artificial plumage – once assembled and fixed to vestments it will be as light and warm as the clothing of birds. Dressed in wings powered by nature, we can transcend our limitations, no longer earthbound like a half-fledged chick constrained to wait for maturity. *With rubber from Brazil*, he writes with a flourish, *man can imitate nature*. You might be forgiven for thinking this is a writer's idle fantasy, so Charles Claude outlines with the precision of an engineer the mechanics of his wings: the rubber arm bands to provide stability, the sockets for each feather in rotation along tubes, a belt around the waist to secure the apparatus, and a breastplate crafted of the same material filled with rarefied air to cushion the flyer against bruises. Nor is this only in his head, for he has roped in his brother to help with the engineering. Each detail is designed not simply for aerial display, but for elegance too. *If a man flies*, he insists, *he must be fashioned to be fit to be looked at*.

So, what are the new things, these good purposes for which his invention could be used? Charles Claude proffers one or two examples of the myriad advantages of detachable wings. Brave men could fly to the rescue of a gallant crew, fast on a sandbank beneath a beetling cliff. No longer would navigators be forced to brave the furious blasts of the Straits of Magellan; they could soar above the storm, charting new routes from the sky. Geographers, geologists and botanists could ascend to the heights of unreachable mountains, their research no longer hampered by natural barriers. Eggs of gulls, down of eider ducks, and multitudinous objects valuable in trade could be gathered more easily. And for those involved in more secretive pursuits, a good flyer could prove an especially good spy.

He sets down his pencil and wipes his brow with a sigh. If he'd only begun working on this idea earlier, perhaps his mother could have taken flight, soaring over the prison walls that unjustly confined her while his father, in betrayal, took a bigamous wife.

Fragments of lies swirl around his head like dust motes. He shakes them off, but still a flush of heat rises to his cheeks. He rustles through the disordered papers on his desk until he finds the title page, his name etched at the top. This was no castle in the sky. With a defiant stroke, he scrawls *Formerly of Crebilly, County Antrim* beneath his name, as if daring anyone to forget where he came from. He lets his thoughts settle before he starts writing again, dismissing the quiet ache of a life that could have been. Then the words come again, steady and clear: *Give me time and the materials, and my wings will own the power of flight.* And what then? We will no longer look enviously at the birds, dreaming of heights we cannot reach. We will rise naturally as the blackbird lifting on the morning light.

His words trail off, not in doubt this time, but in certainty. From his modest terraced home, Charles Claude has done what all great inventors must: harness the weight of life's injustices, coiled tight like a spring inside his chest, and transformed it into the very force that lifts him up through the air, into the future, beyond the glowering clouds on the horizon. There, on the far side of dreams, all things impossible become possible. Though he may not see it, and the world might never understand, Charles Claude Hamilton has already taken flight.

---

Retelling my third-great-grandfather's story here, I recognise the power of perseverance, of daring to believe in possibilities others so easily dismiss – perhaps because I see it reflected in me. While rubber latex, with its exceptional stretch and resilience, appeared to be the perfect material for human wings, it proved, in the end, to be a flight of fancy. Notwithstanding the infeasibility of his invention, I consider the true natural resource in this story to be Charles Claude's elasticity of mind. This, unlike rubber, is something we can all cultivate. His gift for turning hardship into hope, for daring to dream beyond the limits imposed by the blindness of bigots and the prejudice of politicians, is a constant reminder that imagination knows no boundaries.

Untouched by time, untethered by failure, imagination continues to rise, singing as blithely as that blackbird at dawn. And while our family's lost castle was an oft-told tale through my childhood, the true riches left to me surpass any grand estate: a reminder that intangible gifts that lift the spirit and encourage us to dream are the most precious inheritance of all.

## The price of blue

*India • 1860*

In Bengal a few decades later, another kind of flight was taking shape. But this was not one man's invention; it was born on wings of resistance, a collective uprising soaring from the land itself.

---

In the heart of Bengal, the land wears a bruised hue. This is a place where indigo thrives, not in balanced cycles but as an intensive crop, fertilised by the foetid dictates of greed. The *ryots*, the farmers, are chained to their lands by ink rather than irons, their destiny signed away in contracts they cannot understand, in a language that betrays them.

Following the United States Declaration of Independence and the abolition of slavery in the British West Indies, traditional trade routes were blown off course. Now planters in India, mostly entrepreneurs of European descent with ties to the British East India Company, have coerced the locals into growing indigo rather than rice. But indigo's value flows only one way; for the ryots, it is less profitable than other crops – and they can't even eat it! When the ryots sell their first harvest, they don't even earn enough to repay the planters' advance payments, so they borrow more money at high interest rates to sow the next crop, and that's how the cycle of exploitation starts. All the while, the planters watch and smile, white teeth glinting in the sun, like knives.

A many-headed beast, The British East India Company sucks life from the soil, consuming all in its path. The Company already exports tea, cotton, silk, saltpeter, slaves and still it craves more. This blue dye is needed to colour the Indian cotton that fuels the mills across England's industrial heartland. But it isn't long before the land is exhausted, its people hungry, tired, afraid. Like their harvest, the ryots' lives have been compacted into dry cakes of blue, carrying only faint traces of their earlier flourishing. The resentment grows, beginning as a rustling, a secret whispered in the ear while their bodies are bent double in the merciless fields. All it needed was a spark, and for that spark to ignite.

It arrives in the form of a *rubakari*, an official notice, from Ashley Eden, a British official and magistrate with administrative and judicial responsibility in the area. *The ryots are free to decide whether or not to grow indigo. They cannot be compelled.*[10] Under cover of night, the hushed tones become more animated as suppressed longing finds expression in words. If they are free to choose, they choose not to sow the despised crop. In opulent plantation houses, managers scoff at the ryots' audacity and prepare to tighten their grip. Thugs are sent to intimidate, weeding out and punishing the ringleaders. Yet the ryots' resolve is as deep as the pigment that stains their hands. No longer will they bow their heads to serve the English with indigo. The revolt spreads like wildfire, igniting the fields of Chaugacha, Kathgara, Kalaroa, Aurangabad and beyond. The British retaliate with violence and blood stains the land, enriching the soil with the many colours of sacrifice. But these farmers do not meet violence with violence; instead they adopt the principles of non-cooperation, drawing on strengths only the desperate know. *I would rather be killed with bullets than sow indigo*, some declare.[11] Indeed, many are.

By a twist of fate, their collective cry reaches the ears of distant authorities. The Lieutenant Governor of Bengal, returning from an upriver excursion, chooses an alternate route home instead of his usual passage along the Ganges. As he steams towards Calcutta, tens of thousands line the banks: women and men, young and old, gathered from dawn until dusk. What began in the fields has swelled like a river in flood. Beards grow long as barbers refuse planters a shave, dishes omit essential ingredients as traders in bazaars withhold supplies and silver goes unpolished in homes where maids no longer show up for work. It would be folly, the Lieutenant Governor recognises, to ignore such a public display. Planters and farmers alike are summoned to testify before a newformed Indigo Commission.

*Not a chest of indigo reaches England without being stained with human blood*, states Mr Tower, a British magistrate from Faridpur.[12] Meanwhile Reverend Schurr of Kaspadanga issues a stark warning: *The ryots now await Government action, fully expecting redress for their grievances, but they vow that, should they be disappointed, the consequences will be dire. This is not my imagination; I merely convey what I have heard.*[13] After hearing the testimonies of more than one hundred ryots, landowners, planters, missionaries, magistrates and others besides, the Commission delivers their unanimous verdict: it is not the law that is to blame, but the system of coercion. No one will be forced to grow indigo. In desperation, the planters attempt to lure the ryots back with higher wages, offering up to sixty rupees an acre, but the farmers are resolute. *I would sow indigo for nobody, not even my father and mother*, responds Kulin Mundul.[14]

Indigo is a mark of oppression, an emblem of ruin, a curse on their land. Yet the blue hue has birthed more than this, because their suffering has sown seeds of resistance, planted deep in the soil. Watered by blood, sweat, and tears, it will swell into a mighty movement until freedom finally takes root.

---

While the Indigo Revolt did eventually change lives, the oppressive foundation the system was built on, the Bengal Indigo Contracts Act of 1836, has never been repealed, a remnant of colonial rule flapping like a tattered flag in Indian law. Is it not time to remove it from the statute books? Even if this legislative relic is scrapped, this story illustrates how exploitative systems often morph rather than vanish. In much the same way as today's global fashion brands relocate production to regions with cheaper wages and looser regulations, many British planters simply moved to Bihar after the uprising, where exploitation continued. It took Mahatma Gandhi's first campaign of nonviolent resistance to dismantle such practices once and for all. Yet, while some may argue the Indigo Revolt simply shifted the problem elsewhere, this resistance sowed seeds of hope that an exhausted land could once more bear fruit.

And now new hands are tending that hope, this time through a soil revolution that returns to the wisdom of traditional farming. In Erode, Tamil Nadu, Nishanth Chopra leads a movement rooted in a deep respect for the earth and an understanding that soil health is the foundation of agriculture – and ultimately of life itself. At its heart is Ōshadi Collective, a network of farmers and artisans united in a self-sufficient Seed-to-Sew supply chain. Seeds are saved from each harvest to sow the next, freeing farmers from dependency on costly suppliers. Every part of the harvest is valued, nothing is wasted; precious nutrients are returned to nourish the next season's crop. Nishanth's methods reject the interventions of modern farming. Not only are there no harmful pesticides or insecticides, there's no hammering down of prices either. Farmers, not the market, dictate the cost of cotton. Radical indeed.

## Tied up with string

*Mexico • 1869*

A hemisphere away, the long arms of a different empire reached deep into the rocky soils of the Yucatán. This time it was agave, plant of service and sacrament, that bore the weight of colonial ambition. This same plant whose spines the Olmec used in their blood-letting rituals and the Mēxihcah, later called Aztecs by the Spanish, used in their ceremonies would soon be producing ninety per cent of the world's ropes and sacking. And from this green gold the region's hacienda owners, most of whom descended from Spanish settlers, would grow richer than anywhere else on Earth.

---

No longer do the Mēxihcah tear out human hearts, still beating, to appease their god Huitzilopochtli. Now they carve out the heart of their sacred plant, liberating the magical liquid inside. Mezcal is its name, and it's become a source of sweet sustenance for the colonial powers. But it's not just the spirit distilled from agave's severed heart. Each part of its body is claimed as this generous plant bows in service to empire, its medicinal properties catalogued and commodified, its fibres twisted into profit. With leaves as sharp as the blades that fell them, agave is propping up the colonial economy, one plant at a time.

Three hundred years earlier, Francisco Hernández, naturalist and physician, created a veritable encyclopedia of over three thousand plants unknown to Europeans in his *Natural History of New Spain*. Although unpublished for many years, his work was eventually translated from Latin into Spanish and expanded upon by Friar Francisco Ximénez, released in 1615 as *Four Books on the Nature and Virtues of Plants and Animals*. Friar Ximénez recognised agave's myriad gifts, noting: *If people would learn to live in moderation and balance, as is reasonable, this plant would be sufficient to supply all human needs, for the benefits and uses that come from it are almost without limit.*[15] He details agave's role in fencing, roofing, house beams, nails, weaponry, food and drink as well as for fibres to make cordage and clothing, and needles to sew cloth. These early plant chroniclers understood, as Indigenous Peoples still do, that the earth can be a slippery and unstable place where it's all too easy to lose balance, especially when you find yourself in the path of runaway greed. Not only is it immoral to treat nature as a resource to be dominated and exploited, the Friar cautions, it's foolish too, invariably leading to disease and decay. Yet while his books are treasured by those studying native plants of the region, the Spanish quickly turn the pages that speak of moderation.

In the Yucatán's unforgiving soil, few plants grow well, but it is here that henequen thrives. The blue-green spikes of this species of agave stretch across the landscape like a galaxy of shooting stars fallen to the spent earth. In recent years, the landscape has been transformed, and the twin truths of moderation and balance lie buried in the dust. Now, through the ingenuity of José Esteban Solís, who imported a steam engine and converted it to extract fibres from henequen, the bounty swells. This great metal beast swallows the *pencas*, as the needle-sharp leaves are known – up to six thousand a day – stripping fibres from flesh and spitting them out as commodity. Now every hacienda owner wants one. To feed the hungry machine, the land under cultivation doubles, triples, ever more plants, until eventually it increases six-fold. Henequen's cordage crosses oceans, supporting masts, lashing sails, helping ships hold a steady course in the wildest of storms. As twine and sacking, henequen finds its way into fields and factories the world over, binding sheaves of wheat, transporting coffee, sugar, fertiliser, its strength an expression of the resilience of this plant, these people, this land called Yucatán.

Hacienda owners grow ever richer, but they don't want higher wages chipping away at their profits. Their answer is debt peonage, a system that binds men to land with empty promises of freedom, much like the *ryots*. A bitter taste lingers on the lips of the Mayan labourers. This dry crust of land may only yield henequen, but not long ago it was *their* land, *their* henequen. Now it finds an echo in the dry crusts on which they survive. However hard they toil, however little their families eat, the invisible weight of debt bends their backs double from dawn until dusk. Even to fill their tortillas with beans, they're forced to borrow more money from the hacienda owners, heaping debt upon debt. For the Maya, the horizon feels like a far distant dream, and they see little hope of change. The elite of the Yucatán – hacienda owners, traders, investors and bankers – not only hold all the wealth, but the reins of the political and justice systems too, rotating positions among themselves and drafting policies to justify their grip; the outlawed encomienda system lives on under a new name. Nor is this the only rebranding – even henequen has lost its identity, renamed sisal after the Yucatán port from which it travels the world.

As ship after ship departs Sisal's harbour, other boats arrive overflowing with luxuries wrung from the sweat of the land. Friar Ximénez's words reverberate down the years, haunting, persistent. *If people would learn to live in moderation and balance, as is reasonable...* But in this world of excess, moderation is an orphaned virtue, balance is foreign and the only quality deemed reasonable is unbridled ambition. While Parisian fineries pile high within the grand haciendas, outside in their long shadows, agaves rise from the dust, unfettered by the winds of fortune. And all the while, the cogs of industry spin on, powered by shiny new steam engines that strip Mexico's heritage down to a commodity, churning profit for another. With each passing year, the Friar's vision slips a little further into memory.

---

Being an all or nothing kind of person, I've never quite been able to embrace the word moderation or all it implies. Let's face it, that word isn't sexy. But while moderation may lack the thrill of excess, in an age of unchecked consumption, it feels more urgent than ever. This was brought home to me on my last trip to Mexico. I was on my way to Santa Maria del Tule to visit my friend, the weaver and dyer Alfredo Orozco. As the bus sped along on the highway from Oaxaca City, agave sprouted in orderly lines across the landscape, grown not for its fibre but for the lucrative mezcal industry. The state of Oaxaca produces most of the world's mezcal, but as its popularity burgeons, the environmental toll becomes harder to ignore.

A few days earlier, I'd met Vicente Reyes, founder of Hermano Maguey, a community development project that rethinks mezcal's unchecked impact. 'The industry is highly romanticised,' he explained. 'Both the practices and the scale are damaging.' Every year, mezcal production devours around 140 million kilos of agave, leaving mountains of leaves that attract pests and leach toxins. Oaxaca produced fourteen million litres of distilled spirit last year and every litre generates a further ten litres of acidic vinasse, a liquid by-product that can poison the soil and water. While mezcal has become a symbol of Mexican culture, its unchecked expansion threatens the heritage it represents. Vicente is laying a pathway back towards balance. His collaboration with designer Verónica Valdés focuses on recovering *ixtle* – a Nahuatl term for agave fibre – from discarded leaves for artisans who still value the plant's versatility. Because agave's heritage doesn't end at the distillery.

When I arrived at Alfredo's workshop, I found a sanctuary of creativity. For him, excess is found in only two places. The first is on the loom, where he experiments with ixtle, scraps of used clothing, waste paper. There are even banana fibres and a huge indigo-dyed weaving featuring discarded corn husks. The second is in the outpouring of a generous mind. 'There is too much egocentrism and individualism in art,' he tells me. 'Knowledge will die if it isn't shared.' In his gently dazzling way, Alfredo is laying the foundations for a creative community where resources aren't hoarded, but shared.

Through these creative minds, I was able to glimpse a future that avoids replicating the excesses of the past, instead transforming them into something entirely new. This, I believe, is the true counterweight to consumption: an ethic that doesn't simply take but gives back in abundance, especially to those who have been taken from over so many years. Future prosperity lies not in the consumption of heritage but in the renewal of it, not in the commodification of culture but in the eternal strength of community, just as Friar Ximénez saw.

## White rivers

***Brazil • 1870***

As colonial industry tightened its grip on the Yucatán, binding the Maya and their lands with henequen's mighty strands, so too did it extend its reach up the Amazon. For the Munduruku, the arrival of rubber men and loggers represented more than a trade. It was, and still is, an intrusion, a violent force tugging against the roots of their world.

---

Charles Claude's wings come to nothing, but rubber continues to draw visionaries, merchants, opportunists and grafters to Brazil's rain-soaked frontier. The wealth of the rainforest is in high demand. For now, latex flows into rubber boots and waterproof coats, but rubber's capabilities have captured more than my third-great-grandfather's imagination. In London, they've recently laid rubber paving in front of Euston and St Pancras stations to quieten clattering carriages, while in Massachusetts, Charles Goodyear's discovery of vulcanisation has transformed the material's durability, although the true ignition won't come until the next century when the automotive industry takes off and cars burn rubber on roads the world over.

Along Brazil's great river, the Munduruku have long drawn life from the jungle's veins, not only from the *hevea*, or Indian rubber tree, but from euphorbias, breadfruit, cinchona and milkweed. Carefully, they score the bark, working slowly, watching the milky secretions run in pale rivulets down into clay pots, shells or gourds. The Munduruku understand the forest's alchemy, know that as long as they don't take too much, the latex will flow more abundantly, feeding a cycle as regenerative as their relationship with the jungle itself. With practised hands, they work the sap into shape, dipping clay moulds into the liquid again and again, building up layers, allowing each one to dry in the viscous air. Thin layers make bottles and clothing, thicker ones are for boots, sturdy enough to keep their feet dry when the rain cascades in great torrents like the waterfalls where their ancestors live. But it's the small water toys they enjoy most: squirts moulded like cashew apples with a pipe at one end, which they fill up with water. Then, squeeze! These always bring bursts of laughter, a popular prelude to merry making and solemn gatherings alike.

Rubber is not their only material. The Munduruku weave Iço baskets, decorated with annatto red, and knot cotton hammocks for resting in the heat of the day. Fine cotton belts hang low from their waists, strung with the teeth of their enemies. But materials are about so much more than function. While the Munduruku still paint their skin with genipap's blue-black dye, marking themselves in patterns of lineage, there is another covering they like to wear, one that reflects nature in her gaudiest hues. Like the macaws, birds whose plumage announces their arrival as loudly as their insistent calls, the Munduruku love wearing feathers, and the brighter the better. There is only one problem: macaws are virtually impossible to catch, flying straight as a blow dart over the tops of the trees. So, they've learnt to make coloured feathers their own way, rubbing annatto over dull birds' feathers and coaxing them to vivid life with the toxic sheen from poison dart frogs or feeding their parrots on various plants to elicit sunshine-hued feathers that aren't bestowed by mother nature. Colour, though, comes at a cost, because sometimes it's better to blend in than stand out.

The frontier of the forest is moving closer. Streaks of red, brown and black creep across unrelieved green. The Munduruku sense further changes to come, knowing their communal homes lie in the vulnerable borderlands. Other tribes have already been forced into submission by the rubber men, enslaved by empty promises, cheap enticements or the barrel of a gun. And still more arrive: settlers, missionaries, prospectors, cutting clearings in their horizonless forest, filling the void with their flashy goods and foreign god. Camps of rubber tappers or *caucheiros* move ever closer, stripping the trees with an inexhaustible appetite and scarring sacred land.

These newcomers have no interest in Indigenous knowledge; they simply tap the trees dry before cutting them down to make way for new growth. A clash of cultures is gathering and the Munduruku want no part of it. Despite their fearsome reputation, their weapons are no match for Portuguese guns. Nor the diseases that follow in their wake.

The Munduruku slip behind the green curtain, following paths neither authorities nor missionaries know. This jungle is their cathedral, ancient and strong: its towering trees are flying buttresses; its canopy a vaulted roof. Scarlet macaws perch here and there in the greenery, feasting on the fruit of the *tucumã* palm. This is holy ground, where every being draws strength from its neighbour through roots laced like veins. A single body, pulsing with life. The Munduruku world exists on multiple planes, interwoven across space and time: humans, animals, plants and the ever-present spirits of the ancients. One transfigures into another as easily as Karosakaybu, the creator, once turned tucumã fruit into their life-giving river. These hallowed sites, home to the mothers of the animals, demand their protection. Coexistence must prevail over chaos. And so the Munduruku sustain the life that surrounds them, giving and receiving in the unbroken cycle of gift.

Little by little, the caucheiros advance; tree by tree they strip the forest. Deeper they press, draining land of vegetation, draining life from its inhabitants. To them, latex is a crop like any other, to be taken whole leaving nothing behind. Still, they fail to see that if tapped in moderation, these generous trees would give and give more, their creamy lifeblood renewing. Instead, the last tree fallen, the rubber men push on, leaving silence like ghosts where laughter once rippled the green forest crown.

---

Across Brazil, Mexico, India, New Zealand, the nineteenth century's thirst for resources threatened to consume the very essence of cultures that held to a different vision of wealth and possession. The Munduruku recognised that their way of life and their values could not coexist alongside people bent on untrammelled extraction. To the Munduruku way of thinking, wealth flowed from their guardianship of the forest, home of their ancestors, from discerning its rhythms rather than draining it dry. But as latex for their water squirts became harder to find and the echoes of laughter fade from this story, the silence gives space to consider the legacy of our own time.

The pathways of greed lead to emptiness, this much I know, both on a human scale and in our wider world. When we strip the forests, as we continue to do for our clothes and accessories, we leave the soul of the earth emptier, and ourselves along with it. Yet I believe we can change. Indeed, we must. Delaying action increases the risk of crossing ecological thresholds, leading to irreversible damage. Implementing policy and system changes now can help mitigate these risks and shape a more resilient and equitable future. As Margaret Atwood so succinctly put it: it's not climate change, it's everything change[16] – it's about what we value, and what we choose to protect. Ours is an incredibly powerful generation, and the weight of responsibility can often feel overwhelming. But with power comes agency, and the chance to act decisively. So, let's illumine new pathways, propelled by a sense of urgency and empowered to act boldly, as we reconnect, restore and reimagine our common home.

My life here in Branscombe has shown me ways to tread lightly, inhabiting a small space, foraging along the cliffs and shoreline, eating super-locally and, while I may only spend eight months of the year in my caravan, it has provided a blueprint for how I want the rest of my life to be. Yet I know that addressing the climate and biodiversity crisis isn't about retreating, it's about advancing, reimagining how we think, live, act and create. 'We're talking about a global renaissance of systems and society, and that requires expansiveness and outward thinking and interconnectedness,' says Adam Cooper of the climate hope organisation Threads in the Ground.[17] We need wide-ranging solutions that bring us back into union with nature, as the Munduruku well know. Like them, I believe future generations deserve to inherit not the withered veins of once fruitful forests, but rivers of milky-white latex flowing with the promise of renewal.

# PART FIVE
# Shoots of Resistance

## An urgent request
*England • 1892*

The sea is dark, agitated as a cauldron and almost as menacing. Last night, the waves unravelled themselves in fury, frothing and foaming at the foot of the cliff. White caps rolled in from the dawning, a rhythmic deep bass with a crash of cymbals atop. And my caravan trembled.

My daughter, Sienna, has come to stay and we head east along the beach beside dun-coloured sea. Romero runs in circles, eager for us to toss pebbles into the pounding waves. As we walk, we examine fresh limestone crumblings, the margins of our island imperceptibly redefined. The geography of this coast is constantly changing. After the cliff slipped away, my ancestors took to cultivating the newly formed agricultural plats, supplementing their incomes with anything they could grow in this vegetation-rich no-man's land. Branscombe's famously early potatoes perhaps drew their nourishment from the fruitful breezes rolling in from the sea. That and the seaweed. My father has a photograph of Cliffy Gosling, the last of the cliff-farmers, forking bladderwrack into donkey panniers. To fertilise the crop, Cliffy's donkeys would make several journeys a day up steep narrow ways still known as donkey paths. The villagers balanced nature with their own human needs. But what happens when we lose that balance? What sort of beach will future generations inherit when sea levels are predicted to be thirty-five centimetres higher by 2050, a full metre by the end of the century. Will there even be a beach here at all?

We walk as far as Hooken Stacks, twin chalk pinnacles rising into the sky, a craggy remnant of when the land slipped away. This landscape is in constant flux, fashioned by the ebb and flow of the tides just like our world is shaped by the push and pull forces of greed and need. The tongues of the waves lick at these cliffs, biting off chunks of rock, spitting them out, forming ridgeways, reshaping the beach. What is stone today becomes sand tomorrow in an ever-shifting cycle. As I look up at the soaring heights, I'm reminded of my fragility in the face of nature, my ephemeral presence in the passage of time. There's no question that the waves will continue to mould this coastline: the beach a little shorter, the sea a little nearer, retracing the lines on the map. The shoreline management plan for this area calls for no active intervention – a managed retreat they term it, although how we're supposed to manage rising sea levels and falling cliffs, I have yet to find out. Between rock falls and sea level rise, time is ticking for my small and perfect haven. We have little power against coastal erosion.

Or do we? After all, isn't it all connected, just as Humboldt said. Many of these stories illustrate how small actions, when founded on misguided decisions, can unravel entire systems. But if they can unravel, can't they also rewind? And perhaps that's the key. It may not be within our immediate power to prevent the waves reshaping this coastline, but we can still play our part by planting trees and bushes that push down into the earth, allowing the soil to drink its fill. And by fostering our own connection with the natural world, which will, without fail, impact the choices we make. After all, isn't transformation born out of turmoil, out of a pachacuti? When all about us swells dark as a cauldron, perhaps that's where the magic happens.

As my daughter and I stand on the shore looking out to the grey horizon, I understand that the power to shape the future truly lies within our grasp. Like the pebbles I toss into the shallows for Romero to chase, small actions have a ripple effect. Yes, these waters will continue to rise and fall, through spring tides and neap, storms and calm seas, but maybe, just maybe, if we are guided by past wisdoms, if we seek to do good and walk justly, this land will be renewed by the care we take today.

Back at my caravan, I think about how resistance comes in many forms, something this chapter sets out to explore. For some, like Sir William Thiselton-Dyer, it meant challenging modernity. For others, it was about defying the expectations of women's roles. For many, it was about confronting injustice. Yet, not all the stories in this chapter are loud acts of defiance; those have already been told in the history books. Because sometimes the most effective resistance is slow and steady, rooted in the soil of what came before, holding ground against the odds.

---

As inventors and innovators hurtle towards progress, eyes fixed on the next marvel of mechanisation, Sir William Thiselton-Dyer knows that not all knowledge should be hurried. True understanding flows not from theory but practice, from getting one's hands dirty – literally. This is how learning grows, with tender care and in due time, like the plants in the Royal Botanic Gardens at Kew, of which he is director, and the objects he ardently adds to the Museum of Economic Botany. He recognises that these ancestral crafts hold a value beyond the moment, subversive shoots in a world bent on modernity. And if they can be preserved for future knowledge, they might once again flourish, long after the inventions of the day have rusted away.

One morning, Sir William's mind is captivated by a brief article in the *Revue des Sciences Naturelles Appliquées* mentioning how, in some remote hamlets of Languedoc, peasants still make fibre from Spanish Broom, a plant known locally as *genêt d'Espagne*. The craft is now dwindling and in some places nearly extinct.[1] His pulse quickens. Here lies a tradition at risk, and like the plants themselves, such knowledge must be preserved. Specimens of fibre and cloth will have to be secured for the Museum of Economic Botany before the practice vanishes completely. Without delay he pens a letter to the Foreign Office, requesting assistance from the Secretary of State. He assures them that any reasonable costs will be reimbursed, believing this to be a straightforward matter. The Foreign Office writes to their Embassy.

Three months go by before Sir William receives his reply. The Secretary to the British Embassy in Paris has reached out to the President of la Société d'Acclimatation, the publishers of the journal. Regretfully they do not possess such specimens, although the President has written to a correspondent to enquire further. Weeks pass. The Embassy again writes to Sir William, this time enclosing a letter from an official in the Ministry of Agriculture. Monsieur Cornu expresses his deep regret that none of the government departments is familiar with any textiles derived from this plant, tactfully suggesting there might possibly have been some confusion with Sunn Hemp. The letter finds its way back to Sir William.

With a sigh of frustration, Sir William writes once more to the Foreign Office. He acknowledges this matter may seem trivial compared with the pressing affairs of the day, but he insists that it is Kew's mission to study local industries based on plant materials; they might, after all, prove valuable in time. The publication had provided specific information on village locations, otherwise he would not have troubled the Foreign Office and, as for the suggestion of possible confusion with Sunn Hemp from India, well that would be most curious indeed. Sir William emphasises the standing and reputation of the Société. Unless their information is purely fictitious, which seems unlikely, there must be a local industry in France of which little to nothing is known. Perhaps Her Majesty's Consul in Marseilles might investigate?

Sir William is eager to cross the plant off his list, for it's not just Spanish broom he needs for his collection. Next on his list is common milkweed whose seed pods release a silken down that is said to be used elsewhere in France for satin stockings and plush velvet hats. Hops also feature, although extracting the fibres is notoriously troublesome, and rumours persist that some species of mallow produce fibres more beautiful than flax. Sir William understands the cultural value of such fibres alongside the breadth of their utility. Spanish broom is no exception. Homer mentions sparto rope in the *Iliad*, while Pliny describes its use for clothing and footwear. For thousands of years, weavers have crafted this plant into cloth for bags, sails and clothing as well as sturdy cordage; the fibre could rival hemp and linen for fineness and durability. And versatile broom offers more than fibres: its flowers produce a yellow dye, its ashes whiten cloth and its branches are naturally used for sweeping the floor. The bright yellow flowers attract bees, while its shoots sustain sheep and goats through cold winters. Crucially, Spanish broom flourishes on rocky ground where neither flax nor hemp dare grow, its tenacious roots binding scant soil against winds that do their darnedest to sweep vegetation away. It will be a valuable addition to the collection for so many reasons.

In January of the following year, Sir William receives a satisfyingly plump package. The Foreign Office encloses both the requested fibres and articles crafted from Spanish broom, together with an official letter, full of apologies for the delay, explaining the intricate process of turning plant into fibre. It is a painstaking process, distilled over many lifetimes, and it's scarcely surprising that manufacture is now limited to the one village of Lodève in the department of Hérault. Even there, such fibres are now only spun to order. Thread by arduous thread, the industry is dying, As Sir William Thiselton-Dyer examines the fibres, he lets out another sigh – this time with relief. These textiles were no myth. With renewed purpose, he sets about assembling an exhibit for the museum, carefully arranging twigs, fibres in various stages of preparation, yarns, coarse cloths, mattress covers and finely woven sheets. He was just in time to preserve the last remnants of this ancient craft. With the proliferation of cheap cottons flooding the market, the end of traditional textile-making like this seems inevitable, but at least Spanish broom's legacy is secure, preserved within the double-sided bookcases of Kew's singular museum. This knowledge, he knows, is not merely academic, nor are these fibres simply monuments to creativity lost. Perhaps some day, within the evolution of time, these venerable threads will weave their way into the fabric of the future.

---

As it was in Sir William's time, so it remains today: in our pursuit of progress, the imperative to preserve nature's knowledge feels more urgent than ever. Today, Spanish broom fibres are being repurposed, incorporated into biocomposite construction materials to help reduce the building sector's high carbon emissions. These threads, so carefully preserved by Sir William, speak not only of what once was, but of what still might be. Within them lies a reminder that even in an age defined by speed and new technologies, the roots of progress can grow slowly, from traditions that run deep. This truth was evident during my recent visit to Kew's Economic Botany collection, where a white-gloved assistant revealed some of its myriad treasures. These are more than relics of the past. Like Kew Gardens' seed banks, they are brimful of potentiality for our future.

Yet, entwined with the celebrated stories of botanical discoveries is a complex narrative: many of the specimens at Kew Gardens were acquired through deception, exploitation and theft, categorised under the names of their European 'discoverers' in a system that erased any legacy of the true stewards of botanical knowledge. As we move forwards, not only do we need to recognise and reckon with these colonial legacies, but reexamine what progress truly means. Alongside understanding how these fibres can be repurposed for tomorrow's world, we must restore and acknowledge the histories they carry. This is not just about the past; it is about the integrity of our future.

## The changing colour of war

*India • 1896*

Thousands of miles away, Eliza Jessie Stewart was focused on the most prosaic of plants in the material lexicon, of which multiple specimens were almost certainly housed at Kew. Except she had uncovered an entirely different use for it.

---

Eliza Jessie Stewart sees the future of war. Her father-in-law helped quash the 1857 Indian uprising, fighting to secure the British East India Company's stranglehold on trade. Her husband wore the uniform in postings from Afghanistan to Sudan. As for her own father, perhaps the less said the better given that he was court-martialled and his name struck from army rolls. The charges? Embezzlement and conduct unbecoming of an officer and a gentleman. Unlike the men of her family, Eliza has lived her entire life in India, far from her own country's shores. From her privileged vantage point, Eliza can see the Empire reshaping itself, and while she may be a woman, she senses there's a space in this story for writing her name. Because Eliza has seen how the old colours of battle are fading, giving way to the new.

British soldiers used to be walking emblems of empire. Scarlet, blue and white uniforms were designed to blaze like flags on the battlefield, ensuring friends could be distinguished from foes amid the smoke of a thousand muskets. But now weaponry has improved – less smoke, more accuracy – and rifles pierce the air without a haze to hide in, making it easier to pick off a man at a distance. Khaki, from the Hindi word for dusty, has already been adopted by the British army in India, but it's more than the name of this new colour that has been cast through the mire: the dyeing of army uniforms reeks of trial and error. Troops stain white uniforms with cow dung, dirty them with river mud, dye them with tea, coffee, even curry powder in a patchwork of desperation. Such short-lived attempts at camouflage are futile; within two or three washes the colours pale to a filthy white.

Eliza knows there is a better way. She has been watching how the local women dye cloth without any of the usual ingredients: no clay-root or lac, babul bark or dhak and certainly no fugitive turmeric or curry powder. Yet colour clings to the cloth like the last of the day's sun. The answer, she is certain, lies in something almost worthless: the dried burrs left behind after the cotton harvest. Eliza has no idea who first discovered the method of boiling up the prickly remains, decanting, straining and leaving the lacklustre liquid to evaporate in the sun. She doesn't know, and she doesn't much care. What matters is that this grubby residue could be hers to wield, for unlike the troops' patchy efforts, there will be uniformity in her uniforms. These burrs hold the secret to unfading khaki, as tried-and-true as the soldiers who will wear them. More than that, if she acts intelligently, she could recover her family fortune and standing. It would be justice of the strangest kind if cotton were to be the source of her redemption, given that it was one Lieutenant-general Willoughby Cotton, commander-in-chief of the Bombay Army, who presided over her father's disgrace.

As she considers her venture, she feels her father's shadow at her shoulder, reminding her of the consequences of underhand borrowing – how secrets believed silenced can unexpectedly resurface in a pool of whispers. But surely, this is different? This is local knowledge not hard currency, unclaimed and unsung, and to whom do such traditions truly belong? Who'd even written the recipe down until she came along? She files for a patent in London, relating to *improvements in dyeing of a certain fast shade known as khaki.*[2] English patent number 25581 is more than a number: it is her protection against obscurity. So convinced is she of the value of this cotton-waste dye, and so paranoid that others may steal it, she stretches her safety net, submitting patents in India and Australia too.

Eliza has seen the colour of war. In the same year she claims her English patent, the British Army formally adopts khaki as the standard colour for its uniforms in all stations abroad. But their cloth will not be coloured with her cotton-burr brew. Frederick Gatty, an inventor working in the Lancashire mills, has seen khaki's promise too. Inspired by his own trip to India, he has perfected a recipe hewn from the earth; chromium and iron salts produce his dusty hue. Not only does he hold the patent, he has a contract to supply the entire British army.

Eliza Jessie Stewart glimpsed the future of war in all its faded hues. But her khaki will never outfit a single soldier, and her dreams of atonement will be lost to the dust. Gatty's shade steals a march on history.

---

As the century drew to a close, the race to catalogue and claim the botanical wealth of distant lands reached fever pitch. While piracy had long been outlawed, it continued to be practised in various forms, entwined with plants and empire. Nature was to be harnessed to serve prosperity and power, however this was achieved. When British botanist Sir Henry Wickham spirited seventy thousand rubber seeds out of Brazil to the Botanic Gardens at Kew, his actions were hailed not as smuggling or theft but as advancing scientific progress, a noble pursuit and easily justifiable to secure useful plants for the Empire. The successful seedlings were transported to Malaya, which would eventually overtake Brazil as the world's largest exporter of rubber.

The strands of pillage stretch through time to the present day. While piracy on the high seas is less common than in centuries past (although as any sailor will tell you, it's not to be ignored) today's seats of empire are assembled around mahogany tables in the boardrooms of multinationals, driving global exploitation under the guise of trade. Biopiracy – the theft of knowledge and genetic resources from Indigenous communities to patent or control for private gain – is more common than you may think. Protection mechanisms do exist. The Nagoya Protocol is a legal framework for implementing part of the Convention on Biodiversity. In recognition of the role of Indigenous communities as custodians of biodiversity, it seeks the fair and equitable sharing of benefits arising from genetic resources and traditional knowledge. Yet the agreement is underutilised, and few companies have adopted its wisdom in any meaningful way. Like the pirates of the past, profit still finds its way to the richest shores, leaving a familiar trail of plunder in its wake.

The question of who gets to claim knowledge and whose contributions are erased is hardly new. Eliza's story offers more than an example of intellectual theft. It reveals the evolving role of women in the ordering of empire. As I know from my grandmother's stories of life as a naval wife in Singapore, women were often confined to hosting tea parties and cocktail gatherings that reinforced existing hierarchies. Although her khaki failed to leave a lasting mark, Eliza's resistance to the limited roles afforded to women under imperial ideologies and her desire for public recognition speak to a broader shift in female roles. Long sidelined, women were increasingly asserting their power, challenging the structures that had kept them at the margins of Empire. Soon they would find themselves dismantling the very systems they had tacitly helped build.

## Shifting geometries

*Democratic Republic of Congo • 1902*

As collectors and chancers drew on the botanical knowledge of others, the world was shifting shape in ways even they could not contain. The countries from which plant materials were being taken were beginning to push back. Soon, shoots of resistance would break through, reshaping landscapes in ways no patent was able to restrain. Nations were on the cusp of change, and people long silenced were finding their voice. Those voices were strong, and together they would soar. Now comes their hour.

Just as people discovered their power, fibres too revealed their potency. Not only as cords to lash sails to carry stolen bounty across oceans or as a way to acquire guns, but as a medium of cultural survival. We travel now to the Congo, where artisans weave resistance through every strand.

---

Their hands speak in patterns, their threads sing in opposition; black shifts to yellow and yellow returns to black. Here, in the tropical rainforests that green the heart of Africa, fibre becomes artistry, the sinews of culture, in a story that begins long before white men hack a hole through the horizon.

In the fertile lowlands between the lower Kasai, Lulua and Sankuru rivers lies a flourishing kingdom. It has a professional bureaucracy, a tax system, widespread public services, police and military, a sophisticated legal framework with jury trials and an unwritten constitution. Although these people have no name for themselves, outsiders call them Kuba: the people of the king. Not that they are just one people, for the Kuba are a complex tapestry of ethnic groups and tribes united under one leader, their society a model of strength through diversity. And while the kingdom is well-organised, it is also highly stratified, with royalty and nobles at the top and slaves far beneath. Artisans, however, enjoy an elevated status because the riches of this kingdom stem from more than its clement climate, abundant water and fertile soils; more even than the minerals – gold, diamonds, copper, cobalt, uranium – protected, for now, beneath the verdant pelt of the earth. For the Kuba, it is cloth that represents status and rank. It is their currency as well.

Their craft begins with a towering palm that rustles its shuttlecock fronds at the sun. With the longest leaves in the world, raffia shelters all life beneath, from tiny termites in their miniature mountains to the vegetative nests of silverback gorillas. Harvesting the tree is an art form in itself, shimmying up to the high canopy in search of young leaves, before stripping away the outer layers and drying them in the sun so the colour will pale. These fibres are not spun, nor are they spliced, so the cloth is made to fit the length of the leaves. The Kuba might dye the fibres bright red with *tukula* from the heartwood of the camwood tree or a yellow ochre if they happen across brimstone roots. But whatever the colour, there's always black from river mud to dramatise their shifting geometries.

Men and women collaborate to create the cloth, working within their clan. The men take on the weaving, using simple wooden looms strung overhead, softening the fibres between their fingers as they work. Next comes the women's turn. After pounding the cloth until it's smooth as a cowrie shell, they select their embellishment from an extensive repertoire of embroidery, appliqué, patchwork, cut pile, resist and tie-dye techniques. Patterns emerge, geometric and symbolic, a language of lines and shapes, darkness and light.

But then come the shadows, the grasping fists of foreign rule. Protected by geography, their kingdom has been spared ivory hunters, rubber tappers, missionaries and slavers. Until now. Now a blue flag bearing a gold star flutters high above the land. Congo Free State they call it, although that freedom only extends to those wishing to enter; any who flee will probably die in the process. Like a knife through sacred cloth, the Anglo-Belgian India Rubber Company slashes through the Kuba motherland, gouging deep across the terrain and ripping lives apart, all for the sake of John Boyd Dunlop's bicycle and its inflatable rubber tyres. It turns out the rainforest where the Kuba live is rich with a rare species of wild rubber vine, ripe for plundering. There's one problem with this plant that hangs from the canopy: it can't be tapped like the trees of Brazil. Instead, local men are put to work slashing the vines and smearing the sticky latex all over their bodies. Once the substance hardens, Company officials come along to scrape it off, taking hair and skin along with it in a searing reminder of the Kuba's changed state. Failure to meet quotas brings more inhumane punishment: wives and children routinely suffer the loss of their hands, while torture, floggings and rape are par for the course. Any who refuse to work are executed, their rebel villages razed to the ground.

The private army of Leopold II of Belgium seeks not only to control through violence, but to dismantle Kuba culture, targeting the traditions that have long kept them whole. Like one of Dunlop's bicycle tyres pumped full of air, their society is straining under the pressure. The elite feel power slipping through their fingers. In desperation they cling to what is still within their grasp: their art, their cloth, their identity. Rustling raffia fronds do not quiet threads make. The Kuba's response is of boldness, not submission; the king and his nobles double down on tradition. In open defiance of their oppressors, the Kuba channel power through creation as cloth becomes resistance. Patterns grow more complex, plant colours more vibrant. Each piece is a declaration. We are still here. We are the people of the king. And just to be clear, that is the king who walks among us in his raffia skirt, not the one who sits on a cold metal throne in a far-off land. We make in a different way, just as we see in a different way. Although we may not keep written records, our textiles serve as visible chronicles, a corpus of cloth that conveys more truth than in all your musty libraries. We will not forget.

This elaborate fabric is more than regalia for the king and his elite; these garments are walking rebellion, loud proclamations of Kuba values. The shadows may linger, but raffia uplifts the spirit. The Kuba and their cloth will endure, with hope and resistance dyed into every bold thread. Amid it all, the raffia palms continue to wave their shuttlecock fronds skyward, loud witnesses to the atrocities and to a culture that refuses to be silenced. For in the kingdom of the Kuba, fibres cry, patterns shout and colours weep.

---

The harvesting of the white rubber vine and the extraction of minerals hidden beneath the earth came at an unimaginable cost. My words inadequately convey all the horrors endured by the people of the Congo, but photographs from this era brought me to tears – all to put tyres on our bikes. Yet the natural wealth of the region goes far beyond the riches stripped by colonial powers. Perhaps the most valuable resource for our world today lies in something hidden in plain sight, overlooked by all except the Kuba. Raffia is art and resistance, but the palm is more than just a material resource: these plants are entwined with the future of our planet.

The Congo Basin is home to the world's largest tropical peatlands, playing a critical role in carbon sequestration. As well as storing some twenty-nine billion tons of carbon, these generous ecosystems teem with life, with ten thousand species of flora and fauna calling the swamp forest their home.[3] They are not alone. Indigenous Pygmy communities have practised sustainable agriculture here for centuries and continue to help preserve the forest, but as logging and mining operations escalate, these vital wetlands are under increasing threat. Dry seasons are getting longer, and if this pattern continues it could push the peatlands past a tipping point, causing them to release carbon into the atmosphere rather than capturing it. In October 2024, the Democratic Republic of Congo became the first African nation to formally endorse the creation of an international crime of ecocide as a means of enforcing accountability for widespread or long-term damage.

Beyond legislation, community forestry offers one way forwards, where Indigenous and local communities are placed at the heart of the solution, balancing the needs of the environment and the people who depend on it. This model is not just about saving the forest, because the threads of cultural survival and ecological preservation are not separate; they are like warp and weft, strengthening one another. This interweaving of heritage and sustainability has ignited a cultural renaissance too. Under the guidance of Congolese designer Jess Kilubukila, the Kilubukila Cultural Project is breathing new life into the art of raffia palm weaving, with over sixty female artisans reinterpreting Kuba cloth patterns for today's design world. As in the past, so it remains today: every time a strand of raffia settles into the cloth, supported by dozens of other fibres, it announces the strength of Kuba identity. One people united in cloth.

By recognising the profound links between culture, land and cloth around the world, we can move towards a future where textiles and traditions are re-rooted in environmental justice and cultural respect. And where our collective voice is as bold as the Kuba's patterns.

## Nothing is what it seems

*Mexico • 1906*

Thousands of miles away in the Yucatán, reality was likewise hidden, not beneath the forest canopy, but behind a façade of prosperity. Here, henequen, a species of agave, was fuelling an economic frenzy. Yet even when cloaked in pomp and spectacle, even when those in power do their best to silence, truth finds a way of getting out.

---

Nothing here is what it seems. As the presidential cavalcade rolls into town, a carved image of the rain god Chaac stares down from a high triumphal arch. But this is no deity the Maya would recognise – if indeed any had been invited – for the god that sustains the henequen crop has miraculously sprouted a handlebar moustache, fully covering his fangs. If truth be told, he bears an uncanny resemblance to the Mexican president, Porfirio Díaz, who is about to pay his first visit to the Yucatán.

A cheer goes up as the president passes beneath the arch and arrives in the main square, where a cavalcade of floats celebrate the region's long history, starting with the first ripples across the dawn of the world and ending with the triumph of capitalism. Men and women in peasant dress re-enact the well-rehearsed origin story from the *Popol Vuh*, the Maya *Book of the Community*, although none of these actors are Maya. *This is the root of the Ancient Word*, they chant. *Here we shall create, we shall brocade the ancient stories of the beginning*. And how they brocade, how they elaborate, embellishing sad rags with such a high degree of ornamentation that the truth is near invisible, even to the most discerning eyes. We are open for business, the pageant proclaims to the president and his retinue of two hundred cabinet ministers, governors and diplomats, especially if you're a foreign investor with dollars to spare. After all, who wouldn't want to invest in Yucatán's *oro verde*, our inexhaustible green gold? Yucatán henequen binds and bags the United States' wheat harvest and rigs ships the world over. With so much land inhabited by a handful of *campesinos*, peasant farmers, there's plenty of opportunity here, notwithstanding the acres the governor has already given away in exchange for favours or silence. Of course, Mérida is no stranger to dressing up, and why not when it boasts more millionaires than anywhere else on Earth. White limestone houses arise from the ruins of the ancient city of T'hó, their stones imported from Germany, their roofs coated with French terracotta, their floor tiles arranged like a Moorish carpet. Gardens are French or Italianate, all with a fountain. Education is from England. And fashions, naturally, come only from Paris.

Once these festivities are over, the president and his men are whisked away on newly laid train tracks to visit one of the thousand henequen haciendas across the peninsula. Although this isn't any old hacienda. Chunchucmil made headlines when worker Felipe Juárez claimed he hadn't been paid. Said he'd been whipped after trying to escape, had bitter orange rubbed into his wounds and been threatened with conscription if he took matters further. His tale spread as fast as the region's new train tracks – not that it was news to thousands in the same position. Lies, all lies, plantation owner Rafael Peón assured the president as hacienda owners closed ranks. Don Peón always has money on hand to help all misery and misfortune, his neighbours attested, although they told a different story behind closed doors. Where did that fool Juaréz think he was going anyway? they asked, laughing. As an indentured labourer, his only way out was to find another hacendado willing to purchase his debt, and no one was going to hire a *gusano*, a worm who'd run to the press. Even the state governor defended Peón as the most exemplary of employers – the governor did, after all, own the largest henequen export house. Nonetheless, it has been hard to quell the rumblings of slavery across the peninsula and everyone is getting a bit jittery. But President Díaz is a good man, the hacienda owners agree. Hasn't he suppressed the Yaqui rebels in the north and the subversive Maya closer to home? With political power entwined with economic prosperity, stability in the Yucatán is in everyone's interest. And what better place to dispel these unwarranted attacks than the hacienda where the pernicious rumours began?

As the carriages steam by, workers in spotless white clothes raise full moon sombrero heads. Some lift a hand to greet the train, others wave Mexican flags, while women and girls scatter flowers along the route. Porfirio Díaz settles back in his seat, contentment on his lips. He does not see the real workers in the distance, their silence louder than applause. When the train reaches the hacienda, Peón takes the President on a tour, pointing out the fine yellow fibres strung out to dry in the sun before steering him towards the workers' accommodation, bright and clean as the new clothes they wear. The roofs of the cottages are freshly thatched, their interiors decorated with imported bentwood furniture. There's even a sewing machine in the corner. After a lavish luncheon, Porfirio Díaz rises to his feet and the room falls to an expectant hush. He speaks of the magnificent improvements he has witnessed in Mérida, culminating in this highly satisfactory visit. He has long harboured a conviction of the falsity of these accusations and now, more than ever, he is convinced they are no less than slander. The president gets no further, his voice drowned in a burst of applause. But the ultimate act of concealment is still to come: a total eclipse of the moon leads some to suspect even the gods are in on the act.

As the dust settles behind the departing cavalcade, the smiles vanish. Nothing here is what it seems. Clothes, furniture, curtains, even the sewing machines are sent back to Mérida where the triumphal arch is already being dismantled. The actors are swiftly replaced by real workers, day labourers from Salta, Salamanca, Shanghai and Seoul, men and women who labour for half a peso a day; all except the children, who work for free. All are peons in debt bondage, and because of the way the system works, none will ever be able to repay what they owe. Few even understand the wording on their contracts, but all are acquainted with the language of the whip.

Yes, now that the president has left, things can go back to how they have always been, except the hacienda owners can breathe a little more easily. So blinded are they to reality, so deaf to the cries, they fail to hear the hushed voices circulating another truth. While the voice of Porfirio Díaz rang out with assurance, outside in the still air of evening, resistance was taking flight like the four million bats that stream nightly out of their cave in Calakmul. Copies of *Regeneración*, a newspaper fighting for the rights of the oppressed, flutter from hand to hand. The henequen workers know they cannot go on strike, not yet at least. But they listen. They remember. They wait.

---

The story of Porfirio Díaz's visit to Yucatán reminds us that true prosperity is not found in spectacle, but in material improvements to the lives of the poorest in society. As I sit in my cliffside caravan overlooking Branscombe's rugged coast, I think about my own textile legacy. A hundred or so years ago, in whitewashed thatched cottages up and down this valley, my foremothers spent their days with lacemakers' pillows and a set of bobbins on their laps. Lace was once a valuable commodity through Britain and beyond, with inevitable consequences: between 1674 and 1913, the Old Bailey records some two and a half thousand crimes involving lace.[4]

Long before industrialisation sent manufacturers abroad in search of cheap labour, textile supply chains had already unravelled into the dark orbit of outsourced labour. From as far back as the Middle Ages, most manufacturing in Europe relied on the putting-out system whereby raw materials were sent to home-workers or small workshops where workers were paid per item. As well as saving on overheads, this allowed business owners to circumvent the powerful craft guilds. According to *The Lacemakers' Memorandum* of 1698, over a hundred of the village's six hundred residents were employed in the craft, with the local region producing over three million inches of lace a year. Although an inch of fine threadwork could take a full day, pay was a respectable five shillings a week and the weekly West Country coachload of lace was eagerly awaited by London dealers.[5]

The good times didn't last long. As machine netting brushed away this cobweb-like artistry, wages plummeted. The Children's Employment Commission 1862 report notes that *children are put to learn it at a very early age, six being thought the best by some teachers, though many begin at five or even younger. For this purpose they usually go to work at a school kept by a woman in her cottage.*[6] School was a euphemistic term. The report found that only a few minutes a day were devoted to reading in the building's pantry-like rooms, either without a fireplace or blocked up to prevent draughts. In winter, dishes filled with embers or smouldering charcoal were placed beneath the girls' skirts in order to keep them warm without smoke damaging the precious lace.

Meanwhile babies' cries were soothed with tonics laced with alcohol and opium tincture, allowing mothers and sisters to work undisturbed.

Quick to find further ways to profit, lace merchants began keeping general stores where consumption became intertwined with craft. Workers were paid in cider or overpriced groceries instead of cash, a system that effectively cut their earnings and bound them to their employer through continual debt. Although theoretically banned by English law in the early nineteenth century, the truck system remained standard practice for decades. In Devon, the Tucker family became the biggest lace dealers of the area with up to five hundred workers, controlling the whole production from flax through to sale. My father's cousin Betty, born in 1925, told me how her granny used to take their weekly lace delivery across the cliffs to Beer, receiving tea and sugar in return.

Around the world today, families in rural villages still grasp at flimsy threads to raise themselves out of poverty, but fragile young girls often discover the promised opportunities to be as full of holes as my foremothers' lacework. After years leading Fashion Revolution, I came to see clearly how exploitation thrives in hidden places. Just as President Díaz's visit underscored the need for tangible progress over hollow displays, those who benefit from the industry's enduring inequalities must take the lead in shifting away from performative gestures to confront the systemic falsehoods underpinning its flawed value system. This will entail a dismantling of many of the practices that have become normalised, from sourcing and production through to waste, rebuilding new systems from the ground up. Committing to this transformation is the only way to align progress with prosperity.

But this is only one strand. No transformation can succeed without tackling the fleeting nature of what we consume. Were William Morris alive today, I shudder to think what commentary he would offer on the insolent excess of today's fashions. In the relentless pursuit of novelty, the superficial reigns supreme. As Morris understood, beauty without integrity is a shallow illusion, as the workers of the Yucatán would surely agree. Only by confronting those uncomfortable truths we have too long ignored and overcoming our cultural obsession with transience can we prepare fertile ground in which integrity and justice can truly flourish.

Although this may seem an impossible task, I turn to lace as an example. Just as lace is crafted slowly, a single thread twisting into a loop, loops joining together to become a network, anchored by pins until the moment it can float freely, so too can small purposeful actions, united through community, grow a new and resilient fabric for our world, unbound from all that constrains.

## The finest cloth

*Myanmar • 1910*

Lotus fibre is more than the world's rarest cloth. It is the story of a people, a lake and a craft taking root. And it is a testimony to one woman's efforts to honour her faith through the most precious of offerings. As I read of these miraculous threads, with all the elements of a modern-day folk tale, it helped me see that resistance need not be loud; it can settle softly, like lotus roots working their way into the lakebed silt. Nor do traditions need to reach far into the past; they can be birthed in an instant, rising to the surface as something at once enduring and newborn.

---

There is a lake in Myanmar, ringed by marshland and ridged by clouded mountains, where gardens are not anchored in the earth: they float. Across Inle Lake's shallow waters, thickets of vegetation, living and dead, are knit by nature's currents into tangled mats of bamboo, water hyacinth and reed. The people who live here call themselves Intha, or Children of the Lake, and while their houses cluster along the shoreline, with monasteries and pagodas perched on the promontories, their gardens float freely, with no terrestrial limitation save the odd bamboo stakes. Tomatoes, watermelons, gourds and betel vines rise above the ripples. The Intha haven't always been an amphibious people – they fled north seven hundred years ago following the collapse of the Pagan Kingdom – but they are at home on the water. Balancing on one leg as their canoes glide across the lake, the other foot hooked around a paddle, their hands are poised to spear any passing fish.

The early morning mist tumbles down the mountainsides and hovers over the surface of the lake. Daw Sa U, Madam Sparrow's Egg, is on her way to the temple. She stops to gather an offering, a lotus flower pink as the dawn. As she snaps the flower from its stem and mouths a prayer, she notices thin, silken strands clinging to her fingers, soft and translucent in their lake-washed form. Perhaps she has seen these fibres a thousand times before, but today they wind themselves not only around her hands but into her mind. Daw Sa U knows from a lifetime of weaving that such threads could create something rare, worlds apart from the gaudy fabrics the English push on them. Her homeland has been under colonial rule for more than a generation; Manchester cotton blooms in florid colours in the local market, along with Thai silks, traded for opium, their most valuable currency. Her fabric will not be like that gaudy cloth; it will be an offering unlike any before.

Weeks pass. The work is too much for her alone, so Daw Sa U hires labourers to gather the lotus stalks, making sure they leave offerings of coconut and banana for the spirits. Each stem must be prepared carefully, the muddy part sliced away, sharp thorns brushed off with a coarse cloth before snapping and twisting the stems to extract the moon-bright filaments within. It's important not to gather too many each day though, because Daw Sa U has already learnt that the day's harvest must be used before nightfall or risk it becoming too dry to weave. Thousands of stems are collected, and each day she watches her cloth grow inch by inch in a labour of devotion. This fibre, this cloth, flowing in silk-like ripples, seems to carry the lake's very essence within, reflecting light like water.

This robe she is weaving is not for herself: it is an offering for the temple monks. But she already knows she is crafting more than a garment. In the quiet of the morning, as the fibres meet the loom, she knows it's a prayer she is weaving: a prayer to the lake and the invisible threads that have fastened their lives to this place; a prayer of gratitude that she is a child of the lake. And yet it is more than this: it is a garment of impossibilities. Something that was never meant to exist.

At last, she presents the robe to the head abbot of Golden Peacock Hill. The monk feels its fine texture and is lost in wonderment. When he at last finds words, they are to bestow a new name upon her. No longer a sparrow, she will be called Madam Lotus Egg. Daw Kyar U returns to her hut, a new name on her lips and a new garment on her mind, for her fingers are already restless. As well as creating lotus robes for the rest of the monks, she wants to clothe the Buddha statue at the lake's most sacred shrine. She winds the weft onto a stick and begins her next piece of cloth.

It isn't long before other women in her family join her, sitting on rugs spread over the hard earth. Daw Kyar U doesn't guard her discovery as some may have done. Instead, she guides her sisters, nieces and daughters through each step, weaving her knowledge into theirs, and in this quiet exchange, she passes on the fire. So, when Daw Kyar U dies, her family refuse to let the bright flames of her craft fade to ashes. They take up her work, forming a cooperative to keep lotus cloth alive.

In these shared acts and through the rhythm of their looms, a gentle resistance finds form. These women are not rich, nor are they powerful, but they have hands and a lake that provides. With such gifts and their looms, they can create a fabric more noble than any Thai silk, more splendid than Manchester cotton. Perhaps it will be the finest cloth the world has ever seen.

---

There is a saying in Myanmar that lotuses are at their most beautiful when the water is high. These otherworldly blooms flourish when the rains come, purifying murky waters and transforming challenging conditions into nourishment for future growth. Lotus root deeply, surviving whatever the weather throws at them: floods, droughts, even the Ice Age when many other plants became extinct. Even when these plants become untethered, their seeds can survive for thousands of years. Not only are they survivors, lotus positively flourish when they grow back, which is probably a good thing when it takes two hundred thousand stems to make a single robe. And while such cloth may shimmer only in Buddhist temples and the wealthiest wardrobes, this tenacious plant offers lessons for all.

Because we, too, need to cultivate a diversity of approaches that respond to our rising waters. We need deep roots, strong interconnected solutions – economic, technological, community-based and ecological – to adapt to our changing surroundings. And we must create backup strategies for future growth, which, like dormant lotus flower seeds, may seem unnecessary right now, but are filled with latent potential, ready to spring into life when needed. All of us have different strengths and, by approaching the biodiversity and climate crisis in a range of different ways, I believe we can do more than survive. When the skies above are heavy with uncertainty and storm clouds are gathering, we can plant our feet firmly, knowing we too, like the lotus, have found ways to thrive.

## A misnomer

***Ecuador • 1912***

While Daw Kyar U loomed threads of devotion, another tale of tenacity was unfolding in Ecuador. For Eloy Alfaro, the struggle was about more than weaving a beautiful artefact, although that was an undeniable part; it was about creating something firm and unfading for Ecuador's poor – a vision that would not be easily extinguished.

---

Eloy Alfaro is dead. The man who fought for the people has been killed by the people; the man who abolished the death penalty has been killed by a mob baying for blood. Death to the Indian Alfaro, cry the soldiers who break into the Quito jail. A shot rings out and the birds of the city take flight. Minutes later, Eloy Alfaro's lifeless body is dragged through the streets. This time the Conservatives and Catholics are taking no chances: like a pervasive weed, the man they call The Old Fighter has a nasty habit of reemerging with renewed vigour each time they trample him down. Soon the air is thick with smoke from the bonfire of bones that blazes in El Ejido at the northern reaches of the city. But try as they may, they cannot extirpate his legacy. Like the plant that first sparked Alfaro's revolutionary spirit, his ideas have taken root and will continue to flourish, despite every effort to trample its intrepid seedlings down.

It was Eloy's father, Manuel, who spotted the potential of the plant known locally as the *toquilla* or *jipijapa* palm. After hot-footing it out of Spain as a political exile, he married Natividad, a local Ecuadorian, and settled near the coast in the small town of Montecristi. Here, he came across a palm-like plant that had been woven into daily life for thousands of years: the locals used the leaves for roofing and bags, but mostly they wove fine hats to shade their heads from the fierce equatorial sun. Ever since the conquistadors had spotted these tall brimless creations, they had been known as *sombreros de paja toquilla* after the *toque* style popular in Spain. The term had stuck, for the moment at least. It wouldn't be long before they became known by a different name entirely, one that again bore no relation to their place of origin. Far to the north in California, the Gold Rush had begun and workers were dying from heat. Manuel put two and two together and set to work, organising production of the lightweight hats the locals wore from field to finished product. He sent a first shipment to Panama, a popular staging point for eager prospectors making their way north. The hats were an immediate success. By the time Eloy was born, the family had become the biggest landowners in the region, amassing a fortune out of straw.

Eloy was their fifth child, the wild one, the favourite, always off playing in the woods. His friends were the machete-wielding agricultural workers, labourers who cropped the toquilla plant that made his parents rich. Young Eloy saw how the palm-like plant grew in the cloud forest between the Andean foothills and the dry alluvial plain, not in neat plantations like soldiers standing to attention, but amid joyful diversity – trees, ferns, philodendrons, epiphytes, bromeliads, all scrambling for light. His favourites were the walking trees, wondrous leafy creatures that heaved their trunks across the earth, moving towards sunlight or fertile ground on leggy aerial roots. Glistening-Green Tanagers flitted through the canopy in a continual quest for fruit or hung upside down, bat-like, on mossy branches waiting for an insect to pass by their beaks. Monkeys leapt from tree to tree, tarantulas hid beneath the pelt of dead leaves and the occasional armadillo snuffled by. Paja toquilla needed no watering or feeding, only the right habitat in which to thrive. And this it did, so much so it could be cropped once a month and still produce new shoots for the next hundred years.

Childhood couldn't last for ever. As Eloy grew older, Manuel taught his son the hat business, taking him on trips abroad to seal lucrative deals. But instead of giving him a nose for commerce, it opened Eloy's eyes to politics, to the enslavement of Ecuador's people under the dictatorship. And it equipped him with the negotiating skills to do something about it. The name Eloy Alfaro became synonymous with the Liberal Revolution. With such dangerous beliefs, he made a lot of enemies and was always in and out of exile, yet for every foe he gained a friend. As a champion for the rights of the dispossessed, Eloy represented the hope of the masses who twice swept him to power. Hacienda owners' forced labour practices were abolished, day labourers received contracts, land was redistributed. Then he went further, shaking up society by separating Church and State, introducing civil marriages and secularising education. Why not open the doors of universities to women, so they can dedicate themselves to the study of science? he dared ask after seeing women being denied access to higher education. Once his revolutionary mind began whirring, he set about reforming the political system because if women were educated, surely they should take up positions of power?

Eloy Alfaro dedicated his life to securing rights for those who lived on the margins, his liberal dream rooted in the irrepressible plant that symbolised his ideals, and powered by the hands of its workers and weavers. Even after his death, the link between politics and the Panama hat, as it would soon come to be known, was not severed; before long, it was trimming the heads of democrats and dictators the world over. Yet the universality of Ecuador's famous hat signifies more than this: it is a reminder of how revolutionary ideas can flourish across borders.

---

Eighty years after Alfaro paid the ultimate price for his dreams, I found myself in a country still burdened by inequality. My travels there had begun two years earlier when I was carrying out research for a masters in Native American Studies. Returning to work with some of the community groups I had met, I discovered the weavers who crafted the iconic Panama hat, a symbol of luxury the world over, were struggling to survive. Exploitative systems drained profits from rural artisans and placed them in the greased palms of middlemen instead. Some faced more direct threats, with at least two cooperatives I'd interviewed experiencing arson attacks. I decided to create a knitwear collection to support artisan communities before commencing my PhD, but what began as a summer holiday project quickly absorbed all my energies, and watching families send their children to school for the first time reshaped my priorities. I gave up my planned PhD and chose to create change instead.

That was in 1992, five hundred years after Columbus arrived in the Americas, and I felt the weight of the colonial legacy. Choosing to name my enterprise Pachacuti, Quechua for world upside down or time of transformation, was a declaration of my commitment to overturning the colonial narrative and ensuring wealth and recognition flowed into the communities who deserved it. Instead of perpetuating an unequal exchange, I envisioned craft as both a celebration of heritage and a means of empowerment through the power of Fair Trade. It wasn't long before the Panama hat became my metaphor for change, a way to embed sustainable harvesting, social justice and mutual respect into an item of high craft.

At the heart of Pachacuti was a radical idea: reimagining supply chains not as extractive but as relationships built on trust, respect and complete transparency. By working directly with cooperatives of weavers and growers, we traced every hat's journey, from the *Carludovica palmata* that provided its fibres, grown on community-owned, biodiverse plantations, to the final handcrafted product.[7] We created benchmarks rooted in real costs of living. We reinvested in local communities, providing training, education, pensions and healthcare, supporting our artisans in reclaiming dignity and autonomy. Eventually, we became the world's first Fair Trade Certified company. More than just a product label, it was an emblem of our commitment to restore power to the makers' hands.

Even as I stepped away in 2013 to grow Fashion Revolution, an idea born in the bathtub a few days after the Rana Plaza factory collapse that galvanised millions to demand justice and transparency, Pachacuti held to its mission, with my husband, Mark, at the helm. Today, more than three decades since its inception, the brand continues to support artisans in reclaiming their rightful place in the global economy. Because sometimes it's only by turning the world upside down that you can truly set it right again.

## Another misnomer

***Philippines • 1944***

On the island of Mindanao in the Philippines, another tale of trade and resistance was unfolding. Abaca, an inedible cousin of the plantain, grew all over the island, the aerial ascent of its leaves rising like wind-tattered banners against the sky. And from that plant grew both the ambitions of empires and the moral fibre of those who resisted their advance.

---

It did not grow in Manila, nor was it hemp, but when Lieutenant John White shipped a small bundle home, curious about the fibre's potential for rope-making, he labelled it with the name of the port from which it sailed. And as hemp was then the material of choice for ships in the United States Navy, that's what he wrote on the packaging. This was over a century ago, long before the Philippines passed from Spanish to American hands. Since then, the islands have become a flourishing outpost of empire, and nowhere more so than the far-flung trading port of Davao on the island of Mindanao.

Davao was a crossroads of worlds where the ambitions of empires clashed like cymbals and the sea delivered opportunities as readily as a cyclone. The city owed its rise to the crop that bound the world's industry and sent its ships out to sea, the one known as Manila hemp. For years, Americans and Japanese wrestled over this strategic material of war, but it was the Japanese who prevailed, mastering the art of cultivation and paying a better price than their rivals. By 1934, they controlled most of the abaca trade, producing half a million bales every year. For the local farmers, who knew the plant was not hemp but a species of banana, it was a root to hold them steady in the shifting soils of tenancy.

Then came war. Japanese boots marched louder on Filipino soil. The commander-in-chief of the Imperial Forces announced that Japan had no desire for conquest, no designs on foreign territory. But if you obstruct our military operations, we shall crush you with our might, he warned. His words still hung in the air when Davao fell to the Japanese, who claimed the city's abaca warehouses among other spoils of war.

Now up in the hills, resistance is stirring, and it's not just Filipino farmers and Indigenous Lumad who are rising. They have been joined by those Americans who, refusing flight or surrender, have taken to the hills too. Among them is mining engineer Colonel Wendell Fertig, who admires the locals' spirit but despairs at their disorder – seeds of rebellion are scattered across the highlands in three hundred or more rebel groups. Fertig promotes himself to Brigadier General. Under his command a cohesive force arises tens of thousands strong, fighting with weapons smuggled in by plane and submarine. They draw strength from the soil too; abaca becomes a tool of war, twisted against those who had not long since bought the entire crop. Abaca nets camouflage hidden camps; draped over tents and equipment, they blend with the understory, tricking sight from above. As Japanese planes search for signs of life, they see only forest, yet beneath this botanical veil the resistance hums with activity: ambushes and sabotage are planned, weapons cleaned, wounds tended, intelligence sifted. Of course, the Japanese come by land as well, so the guerrillas loop abaca into snares that snatch soldiers' feet from beneath them as they trample the forest floor, and nooses that tighten around necks with merciless precision. The jungle trails once used to haul abaca to market become a labyrinth of death for the Japanese. The banana trees, allies to the resistance, swallow their screams.

While men reclaim the jungle, women have their own battles to fight. Their hands move deftly as they weave abaca baskets, sturdy containers that appear innocent enough. But these are no ordinary vegetable wares. Brimming with fresh produce for market, the baskets' false bottoms conceal bullets, medicines, messages, maps. Eyes ahead, baskets clasped tight, they make their way through the throng under the feverish sun, threading resistance across enemy lines. In their capable hands, everyday commerce becomes an act of defiance.

As the tide of war starts to turn, the guerrilla force of Mindanao becomes indispensable to the allies. Farmers turned fighters, weavers turned warriors guide American soldiers through the treacherous terrain, their knowledge of the land turning each winding trail into a strategic advantage. The Japanese grip on Davao weakens and finally breaks, leaving a million or more dead, cities in ruins, plantations neglected, communications destroyed. Rebuilding is a priority for the newly independent nation, and restoring production of the world's strongest fibre will play an important role in their strategy. That's when the people of the hills start to wonder whether this fibre that bound them to empires, that wove the fabric of resistance, could perhaps craft a new story of liberation?

---

Abaca no longer rigs ships the world over, but the fibre maintains its bond between Mindanao's people and their land. Although efforts to revitalise Mindanao's abaca industry progressed slowly in the post-war years, today the outlook is brighter. On the same hills where resistance took root, small-scale farmers harvest the plant not from monocrop plantations but from intercropped fields where it thrives alongside coconut palms and native trees, restoring biodiversity to lands stripped bare by empire. Its extensive root systems bind the soil, preventing erosion and landslides, and countering the impacts of climate-related disasters. Abaca thrives in intense heat and prolonged rainfall, delivering good yields time and again.

When I visited Fashion Revolution's team in the Philippines a few years ago, I saw how abaca continues to sustain rural livelihoods, crafted into accessories and homewares for local and global markets. Returning home, my suitcase was stuffed with abaca gifts for friends and family. Beyond its use for accessories and emerging applications like biocomposites, Manila hemp is an unsung thread through our daily lives. If you're reading this with a hot drink to hand, chances are your teabag or coffee filter was made from abaca – holding things together, just as it always has.

## Finding strength in weakness

***Bangladesh • 1952***

Bangladesh has long been caught in the crosscurrents of history. The fortunes of East Bengal were reshaped first by the Partition of India in 1947, then by its transformation into East Pakistan and later by the birth of the independent nation of Bangladesh. Before Partition, the region's fertile lands and labyrinth of rivers sustained a thriving jute trade, with the river port of Narayanganj connecting the jute grown in Bengal's hinterlands to Calcutta's mills and the global market. But Partition split Bengal in two – East Bengal for Pakistan, West Bengal for India – and the rivers that had carried a seamless flow of jute to the mills now divided two nations. Narayanganj's farmers and traders found themselves navigating a fragmented reality.

---

The rains are at an end. In the glistening fields, the jute has grown tall and is quivering in readiness. Farmers wade through waterlogged fields, hands slick with mud as they slice the jute plants at the base and tie them in bundles. The work is slow, the air humid, the bundles heavy. And this is just the start. Soon, the stems will be plunged into the river to soak, allowing the soft inner part to rot away. Retting is a slow and pungent process, but essential for exposing the plant's golden fibres. After weeks in the river, the softened stalks are pounded and laid out in the sun. It is a laborious process to extract value from this rain-soaked land.

At the end of the day, jute still isn't perfect. Although stronger than cotton, it won't withstand a good tugging, unlike henequen or hemp. Yet where tougher fibres might falter, jute finds strength in a measure of frailty. It is ideal in the garden for tying wayward stems to a stake without damaging the plant. Likewise, the British and American postal services prefer jute over sturdier fibres as it can tie letters without harming the envelopes and yet be snapped by a man's hand. Those letters are then transported in jute canvas sacks tied up with jute string, and sometimes their paper might even be made from the stuff! It's a curious thing that what's seen as a weakness can be exactly what's needed. Much like the peasant farmers who harvest the jute. They too have been twisted and pulled in all directions, but still they do not snap.

Not long ago, their lives were very different. Jute was a crop of promise, willingly sown. With its shifting soils and annual floods, this silt-rich region produced three-quarters of the world's jute, and unlike indigo, the farmers needed no colonial coercion to grow it, no unintelligible contracts, nor inescapable debt. Their crop travelled down the veins of the river into the heart of the delta, the port of Calcutta, from where it was shipped to the world. Jute made them rich – well perhaps not quite rich, but it gave them a taste of what that might feel like. Yes, men still tied their lungis and women draped their sarees, but now they had a choice: to make them from the Manchester cottons that filled the bazaars and melas or from locally woven cotton and silk. And when the monsoon rains came, their clothes remained dry under smart English umbrellas. Corrugated iron roofs slowly replaced thatch, although the walls were still bamboo and there was still just the one room. Inside, a kerosene lamp would flicker to life as dusk fell, lit with safety matches from Sweden instead of those unreliable British brands.

Jute turned peasant farmers into consumers and the markets, filled with imported luxuries, gave voice to their growing desires. Yet they bought more than this. Through educating their sons, they could purchase the power to break free from subsistence farming and move into town to join the professional class. Jute reshaped their future. Not that the British administrators recognised this as modernity – they were looking for petticoats and partitioned houses and saw such extravagances as another sign of the foolishness of the peasant class. But their opinion was of little concern to those growing wealthy on jute. Manchester cloth and all the other trimmings were only tools, a way of carving out a life that wasn't defined by the boundaries of a field. Modernity wasn't about mimicry; it was about having a choice.

Having a choice. That's what brought them here, to this place of change, this hollow utopia. Now they question how they could have believed all those electioneering slogans promising a land of justice and truth where the soil would belong to the tillers and where jute would fashion a new economy. They hadn't understood that they were dispensable, entangled in trade negotiations in far distant lands. Then, just as they were bringing their crop to market, the rivers that carried it to Calcutta were cut through by a new border. While the waters still pass freely, trade is no longer seamless; not only are they partitioned from India, but from the mills that process their crop. Beyond their fields, boats still ply the rivers, piled high with jute, but now Narayanganj buzzes with customs officials ensuring no one loads steamers without the necessary paperwork. The river is peppered with checkpoints.

Of course, this brand-new province has a shiny new jute mill to match: Adamjee Jute Mills towers above Narayanganj, the din of its machinery heralding a new era of industry. Opened a year ago, it is the largest of its kind in the world, a solid declaration of independence from India and a symbol of the new nation's aspirations. And this is just the beginning of the new State establishing its sovereignty over their golden fibre: from defence to capital equipment and consumer goods, all have to be financed from the export duties on jute. As if that's not enough, farmers now need a licence to grow it, paying a levy of a rupee an acre. Naturally, they find ways around the form-filling and taxes, and smuggling flourishes for a while, until the government calls in the army with shoot-to-kill orders for any boat or ox cart that dares cross the border under cover of night. Even possessing jute within five miles of the border is an offence. Now every bale of jute represents a different calculation: not how much they can buy with their profits, but between going without and smuggling over the line. They know the risks, but they have tasted the rewards. It's of little consolation that, unlike cotton and indigo, jute's tender leaves can be eaten, making a particularly good curry.

Jute would be their safety net, strong enough to support a nation – that's what they were told. Now they see it's just another fragile web, stretched thin across borders with threads that divide. All the while, as farmers weigh the risks, the jute mills of Narayanganj grind on, churning out yarn from the plant's fibrous stalks like thinly shattered dreams from the rupture of a nation. And in these golden strands, the story of East Bengal is written over again.

---

The resistance of East Bengal's peasant farmers was not marked by violent uprisings like the indigo riots a century earlier. Instead, it manifested in countless small acts of survival, daily struggles against a system that demanded they be producers yet no longer offered a pathway for them to become consumers. The promise of a better life remained just out of reach.

Today, jute remains of material importance to the economy of both modern-day Bangladesh and India's West Bengal, finding its way into sacking, fabrics, hessian, garden twine, rugs and carpet backing. Beyond these traditional uses, waste from jute spinning is an important ingredient in technical textiles, including geotextiles to control soil erosion, and as an affordable ingredient in composites for cars and construction. Deeply woven into the fabric of Bangladesh's economy, while jute's economic importance endures, the nature of prosperity has shifted. From agricultural roots, a new economic life has grown up, not from the soil but from hands and machines.

Has that much really changed? Today, it is still the poorest, the garment workers, who are cogs in the machinations of global capital, their wages and working conditions shaped as much by decisions in distant boardrooms as by local government. Yet amid their ongoing struggle, one thing remains constant: in every stitch, in every seam, their dreams continue to flow like the rivers that run from Narayanganj out to sea, unconstrained by factory walls and unfettered by all that binds. Like jute's gold-spun strands, Bangladeshi garment workers know their dreams aren't held by the strongest of threads, yet they cling to them all the same.

## Seeds of liberty

***El Salvador • 1980***

The history of El Salvador is one of violent eruptions, written in blood and ash. And it is an epic tale of resistance, of a people who clung to the land despite every effort to uproot them.

---

Archbishop Oscar Romero stands in the pulpit addressing the faithful. Mild and obedient, Oscar Romero had no interest in radical politics when the oligarchs appointed him three years ago. But he is no longer the man they thought they could control. Last month he even wrote a letter to Jimmy Carter imploring the US president not to send military aid to El Salvador's government, saying such a move would only sharpen injustice and repression.[8] Oscar Romero has ceased to be the oligarchs' pawn; instead, he is a thorn in their side. In his Sunday sermon at the Sacred Heart

Basilica, broadcast by radio throughout the country, he calls upon the National Guard, the police and the troops to reclaim their consciences and stop killing their sisters and brothers. Any reforms are meaningless if written in blood.[9] Oscar Romero refuses to stay silent, fully aware he might be signing his death warrant. Outside, the air is thick with expectation, as if the earth itself is holding its breath. Nobody wants a repeat of what happened almost fifty years earlier.

On a night in late January 1932, five thousand campesinos roused themselves from their beds, sharpened machetes and creaked open the doors of their huts. Stepping out into the dark, their footsteps cushioned by rich volcanic earth, they wondered if they would ever return. Volcanoes were both a blessing and a curse for the Indigenous Pipil: they created the fertile soil that allowed people to scratch a living from corn and beans they grew in their yards, and they provided the perfect mixture of nutrients for luxury commodities like cotton, indigo, coffee and sugar – the sources of their oppression. Those and the military dictatorship. Anyone seen bearing a machete, wearing traditional dress or speaking Nahuat could be killed on the spot. Then the landowners stole more land. Their land. The Pipil had endured long enough.

As they crept through the cloud forest, a palpable tension hovered over the land. It was picked up, perhaps, by the earth itself, which began throbbing as if it were alive. Mighty Izalco grumbled itself awake. Spewing rock and flame, it wasn't long before the volcano drowned out the noise of the least circumspect of peasants. In an echo of neighbourly solidarity, Fuego volcano in Guatemala popped its top as well. It was a sign, of that they were certain, although whether a portent of success or a warning, none could say. With ash silvering their hair and fire growing in their bellies, in Juayua, Nahuizalco, Izalco and Tacuba, the campesinos headed for the houses and fields of the oligarchy, the supply lines of the military dictatorship. At their head was a woman, known to all as Red Julia, whose name lingers on in legend and song.

The uprising was swift and unexpected, the military and landowners taken unawares. But not for long. Retaliation was merciless. While the rebels killed no more than a hundred, the government sent in the army and militia to slaughter all men over the age of twelve, along with anyone else suspected of communist sympathies. Not that the soldiers had time to differentiate. Women were tortured, bystanders gunned down. So many died, no one kept count: ten thousand? Thirty thousand? Fifty thousand? No time to bury the bodies. Once the uprising had been crushed, the authorities assured the oligarchs no Indians existed in their country anymore. It was a lie of course. The people remember. The land remembers.

The minority continued to control the majority of land. No, not just the land, but water, plants and animals too. The oligarchs had a tight grip on everything. They had given up growing indigo and coffee production was declining; cotton was the future, so they said. But everyone knew what they weren't saying. That the forests, the only ones left standing on the Pacific coast, would be razed to the ground. That the land would be pumped full of pesticides and insecticides, poisoning the people, destroying ecosystems, forever altering the delicate biological balance of the coastal lands. That the cotton fields would not create good jobs, nor safe ones either, with work as wavering and perilous as the toxic tide. That any campesinos who still owned land would be forced to sell it to the oligarchs. That another wave of misery would sweep across the fractured land the Pipil still called home.

Oscar Romero grew up in this segmented world. A cautious man, he rose through the hierarchy of the church as a safe choice, a quiet cleric who supported the institutional church, which backed the oligarchy and the military that enforced their rule. The church, like land, is rarely neutral. But after he was appointed Archbishop, Oscar Romero changed. No, Oscar Romero is changed. By the death squads and their continued killings. By the rape and abuse. By the exploitation of campesinos in a land where the minimum wage is as much a fiction as free elections. By the murder of his Jesuit friend who spoke out against the corruption and injustice. Oscar Romero starts to question why the wealthy and powerful, the very people who supported his election, continue to sanction this violence that sweeps through their land; the violence that keeps them in positions of power. Oscar Romero goes out into the countryside and begins to understand that these stories he's been hearing are not made up. *But how is it possible for good Christian people to do these things?* he asks. *They do that and more!* comes the unflinching reply. *Do you know how these Christians who are such good friends of yours make up for these outrageous tricks? With little Christmas presents. On a certain plantation, where some close friends of yours live, do you know what they gave their workers who picked cotton for them, frying their backs in the hot sun like pork skins? A pair of underwear worth three colones! Three colones is what they saved every day from each one of them by not giving them any food to eat!* [10]

Oscar Romero starts compiling evidence; it soon fills seven dossiers. He carries these stories back to the cathedral and proclaims them from the pulpit, his homilies transformed into indictments of power. Although the oligarchs accuse him of forsaking the gospel for politics, he knows the true church speaks out on behalf of the poor and against all injustice. Of course, he receives death threats himself; once the cotton barons realise he's a danger to power, Oscar Romero knows his days are numbered. *If they kill me*, he tells his friends, *I will rise again in the people of El Salvador. … May my blood be the seed of liberty.*[11]

The day after his Sunday sermon they silence him, a single bullet through the heart as he celebrates mass. The blood that spills on the church floor runs into the soil that hushed the footsteps of Red Julia and her five thousand comrades. The blood mingles with the ashes of Izalco's fury. And it unifies with the sweat of campesinos who toil in the fields. More than a quarter of a million of them arrive in San Salvador for his funeral. It is a seed, as he foretold.

---

Those seeds took time to bear fruit. El Salvador was plunged into a civil war that lasted until 1992, its people terrorised and murdered by the United States-aligned death squads. El Salvador was where Carter, then Reagan, decided to draw the line against communism, framing resistance as a threat to global order while ignoring the blind cruelty inflicted on the ground. Yet from campesinos to curates, El Salvador's people held true to their vision of liberation.

Those were tumultuous years across Central America. They planted an early seed in my mind, which sent me on a long journey to disentangle the ties between land and power, between landlessness and oppression. In the final years of El Salvador's civil war, I was studying for my masters and following the resistance movements from afar. As I read books on liberation theology and saw how the extraction of resources and the cultivation of commodities were entwined with exploitation, I came to understand that liberation is at once collective and deeply personal. Like Oscar Romero's transformation in the cotton fields of El Salvador, liberation means questioning the systems we have inherited, reckoning with injustice and nurturing hope when all around us are slumped in despair. Liberation is tied to place, yet spreads its seeds far beyond any borders, carried on the wind, washed in by the tides. Seeds lay down roots wherever they land, criss-crossing communities and bringing people together in a vast interconnected web.

El Salvador's long civil war is over, but the erosion of its soils, the decimation of native forests and the insecticides poisoning its rivers serve as a reminder that the revolutions to come require more than justice for the oppressed: they demand radical healing of wounded lands, reversing long histories of greed. As we teeter on the brink, it is only by challenging those who profit from destruction that we can ensure true prosperity for all, because the struggle for liberation cannot be separated from the earth itself. The land we walk on, the lands we come from, are all part of this larger story, where each thread reminds us that what happens in one place echoes across the planet.

Throughout this book, plants have been more than a backdrop: they are catalysts for resistance, lifelines for communities and allies in the ongoing fight. I see now that to resist is to care, not only for each other but for the intricate web that binds all life on Earth. These stories act as both mirror and compass. As I have navigated through worlds and time, I have seen how plants heal the land, cure the body and provide a poultice for the soul. Through them, I have learnt that resilience springs not only from the land but from the stories it holds. And in weaving these narratives, I've rewoven my own connection with nature. When I arrived at Undercliff, my caravan, I was worn down and fragile, like the wave-weathered cliffs. I came to Branscombe without knowing what I sought – only that this small patch of earth halfway up a cliff was something I had waited for all my life. Six months on, I feel rooted and refreshed, steadied by the enormity of the sea outside my window and the rhythm of its tides. And when I turn my back on the coastline and walk up through the combe, I can read my family history across the lines of the land.

Now, as this chapter ends, so does my own. I'm packing up my caravan for the winter and preparing to travel to El Salvador and beyond. The day before my departure, I awaken to a rare glassy sea, perfectly still with barely a ripple reaching the shore. The vast silence is punctured only by a peregrine cry. The day is lightless, overshadowed by clouds, but a bright golden thread is drawn across the horizon, like a promise. It hangs there all day, neither dimming nor brightening. A reminder, it seems to me, that every ending carries within it the seed of a new beginning.

# PART SIX
# As Threads Spiral On, So Does Tomorrow

## While our trees gently weep
*Peru • 2024*

Even as our material world hangs by a thread, it is still holding, repairable if we act swiftly while loose strands remain. All over the world, people are rising to this challenge and doing exactly that. Awarded a Churchill Fellowship, I embarked on a journey through Peru, El Salvador, Guatemala and Mexico to explore the unfathomable wisdom of plants, interviewing pioneers, guardians, innovators, artists, scientists and mavericks. Plants offered their own wisdom too, part of an entwined ecosystem upholding each biosphere; nature and humanity as one. This two-ply thread of human and ecological narratives forms the warp and weft of this chapter, helping to reveal how the patterns of past and present might be reimagined for our shared future. The stories in this final chapter are vignettes. Though limited in number, they encapsulate the insights I gathered during a transformative experience immersed in cultures and textile traditions across Latin America.

Peru was the perfect landscape within which to explore the fertile intersection of plant wisdom and cultural traditions. From my very first day, I was immersed in a tapestry of stories that reflected the country's diverse vegetation and the contradictions and complexities that simmer beneath the surface. I danced on the mountaintops. And I cried alongside my interviewees more than once as they shared traumatic stories from within their communities rarely voiced to outsiders. While these do not find their place in this chapter, which is ultimately a hopeful vision, they are part of the same fabric, their threads running invisibly alongside. The botanical abundance of the land and its rich cultural heritage endure alongside its wounds. It is within this fraught coexistence that these seeds of possibility grow.

One of my most memorable encounters was with Sadith and Olinda Silvano from the Shipibo-Konibo community of Cantagallo, Lima, not least because they sang constantly while they worked, infusing their designs with meaning. Watching their artistry, I came to understand how restraint can speak volumes when it comes to patterning. Their intricate tracery unites two principal elements: the background colour of the fabric and a symmetrical pattern of plant-based resin that draws upon elemental pathways in their mythical past. This is not about inventing something entirely new, although that does happen, but about channelling something timeless and universal. The artist envisions pre-existing patterns, overlaying them onto the fabric and shaping them to harmonise with its form. Any unmatched portions seem to dissolve into invisibility, leaving a design that transcends the textile itself. This is more than a pattern; it is a portal, a glimpse into the universe's vast, interconnected web. Through the mirror-like balance of Shipibo-Konibo textiles, I observed how minimal elements can tell the most complex of stories. It is within the lattice of interconnected parts that clarity and meaning emerge.

Inspired by my visit to Cantagallo, I offer these final stories as illustrations of our ability to sing a new future into being. Through their symmetry, I hope they weave a coherent whole, a pattern of hope and a vision of renewal, drawn from the threads of what once was, what is, and what one day could be.

---

While trees are illegally logged to clear land for mining, while rivers are polluted with mercury, while the forest burns, the trees of the Awajún weep, not out of sadness (although, surely, they do if we could but understand) but with a free-flowing white sap. Known to the Awajún as *shiringa* and to others as *hevea* or natural rubber latex, this is the same milky liquid harnessed by Charles Claude Hamilton for his earth-defying wings. Over two thousand plant species contain latex sap, including lettuces and dandelions, with *Hevea brasilensis*, a member of the spurge family, the principal species used for rubber production.

Deforestation and poverty, two relentless forces, are intertwined in the Amazon. Condorcanqui province in the Amazonas region is one of the poorest areas of the country, but in terms of biodiversity, it is among the richest on the planet – for the time being at least. Almost five hundred species of butterfly flit beneath the canopy. Among them moves an Awajún rubber tapper, carefully scoring the bark of a shiringa tree with two or three diagonal slashes, moving on to the next, returning later in the day to collect the latex that has dripped into the cup. He won't take too much and will allow the tree to rest before tapping again. The white liquid gleams in the dappled sunlight. Seven drying cycles will transform it into a supple and lasting material. The Awajún wish to live as their ancestors did, cultivating achiote, cassava, coffee and cacao, catching catfish in clear rivers, breathing pure air, tapping the shiringa trees, migrating from time to time to preserve the balance of the land. Rotating their settlements allows the forest to regenerate, ensuring these resources will sustain future generations.

The story of latex doesn't end with tradition. In the bustling city of Lima, a thousand miles from the rainforest but it feels like half a world, one pioneering company is taking this ancient material and reimagining its potential. Down an unprepossessing alley in a residential district of the city, Caxacori Studio works hand in hand with fifty-nine Awajún families across six communities, blending their plant knowledge with cutting-edge material science. Latex is mixed with waste from Peru's food and drink industries – grape skins, cacao, corn – and coloured with plants like *airampo*, a cactus from the high Andes, or achiote and genipap from the rainforest. These mixtures are used to create bioleathers and soles for shoes. The studio is even exploring adding eucalyptus oil to the soles, imbuing them with antibacterial properties – a small but impactful leap in functional design.

'If people really understood how fashion can destroy or protect life, we would all live better,' says Rosalia Manuig Taa, an Awajún woman who understands the role sustainable latex can play in defending her land.[1] At Caxacori Studio, Rosalia's philosophy is their guiding principle. The studio is more than a business venture; it is a model of collaboration, not just with the Awajún but with local government, private institutions and state organisations that work to protect the Amazon. And it is a shining example of Fair Trade and ecological stewardship. At the time I visited, over seven and a half thousand hectares of forest had been preserved through this project, conserving biodiversity and fostering economic resilience within the community.

---

Global demand for natural rubber is increasing. Many of the world's biggest buyers have joined the Global Platform for Sustainable Natural Rubber whose members commit to twelve principles, helping to prevent deforestation and supporting the restoration of stripped rubber landscapes. It is progress, but change isn't happening fast enough for communities on the bulldozers' front line. (And while the crash of deforestation echoes loudly, the soundless uprooting of other plant life still thrums across the world, with impacts we are yet to discover.)

But just think what could happen if we picked up pace. Imagine walking down a street where the rubber soles of everyone's shoes were made from sustainable natural latex. Imagine if Rosalia's vision materialised, and we all understood how fashion can protect life. Imagine walking more lightly upon this Earth.

## Where coloured cotton grows

***Peru • 2024***

Before embarking on my research journey, I was aware that Peru's native cotton spanned a spectrum far beyond white. Yet while I had come across its beiges and browns, lilacs and greens, I soon discovered there was a deeper story to tell. After visiting Caxacori Studio, my next visit was to the ethnobotany department of Lima's La Molina University, where Gabriel Pihuaycho, one of the research students, showed me a photograph of a shade beyond my imaginings: blue cotton growing in the Amazon. No one knows its exact location. I immediately offered to join any future expeditions to locate it, although this seems unlikely given the department's severe underfunding.

—

Away from the bustle and petrol fumes of Lima, my itinerary carried me east to the Yanesha community of Santa Rosa de Pichanaz on the brow of the selva. It was here I met Vabi Miguel Toribio, a woman in search of a colour lost to all but the memories of the ancients. After completing her schooling, Vabi went to Lima to find work, leaving behind her traditional *cushma* tunic and her mother tongue. At sixteen, she found herself a teenage mother, working day and night as a maid. Invisible in the urban sprawl of the city, she remained a victim of discrimination despite all her efforts. Feeling the loss of her identity, Vabi returned to the rainforest to pick up the scattered strands of her heritage. Back in her community, she discovered few women continued the practice of spinning and weaving, and the coloured cotton plants she remembered growing by the Iscozacín River had mostly disappeared. Vabi, I quickly learn, is not one to be deterred. She taught herself to spin the local cotton and, once she had mastered the art, began teaching others. Then began a new journey of experimentation, this time with colour. Her palette mirrored the rainforest itself: purples, browns, yellows and oranges from plants like *yetsñor*, mango, matico and *payón*, a tall evergreen tree they call the mother of dye plants. Even though payón yields an inky hue when mixed with masato, a paste derived from fermented bitter cassava, Vabi sensed an absence. She visited the elders, who confirmed black cotton had existed: it grew across these hillsides. Some even recalled spinning its threads. Now those plants are gone, or so it seems. But Vabi refuses to let invisibility stand in her way. 'Black cotton must still exist somewhere,' she told me, her voice taut with emotion. 'One day, I'll find it.'

Vabi was not the only woman searching for the lost and overlooked. When I started to plan my research journey, I knew I must travel to the north of Peru to meet Yolanda Contreras. Raised by her grandmother in the dry desert of Mórrope, life was a constant struggle to make ends meet. But one day Yolanda found herself drawn into something far greater than she ever imagined.

---

Some called it courage, others said it was recklessness when Yolanda Contreras gave up her job at the chicken factory, especially since she had a young family to support. It all started with the group of women she passed on the way to work: two o'clock every Tuesday afternoon, always there, deep in discussion by the side of the road. For weeks, Yolanda only glanced at them in curiosity. Then one day, she summoned the nerve to ask what they were doing. That's when she heard about their dream: to revive the native cotton their forebears once grew, the cotton that sprouted brown and beige, yellow and orange, lilac and green – and perhaps black and blue. The plant they called simply *algodón El País*: country cotton. Yolanda felt an invisible thread tug her back to her grandmother, who had made clothes from seeds passed down like heirlooms. Yet it wasn't just nostalgia that convinced Yolanda to join them. Coloured cotton was an opportunity, a means of escaping the drudgery of the chicken factory where every dead bird was docked from your pay. Yolanda hung up her apron and joined their crusade.

There was only one obstacle with growing such cotton – well, two or three it eventually turned out. The most immediate was finding seeds. The women searched high and low throughout Mórrope, scouring every ditch and hedgerow, combing their homes and fields. Eventually they found a handful of seeds buried in the fluff of mattresses and pillows, and a few forgotten plants clinging to life in the countryside. Starting with so little felt like an impossible task, particularly as Yolanda had no idea how to cultivate them. The region has a desert climate and, with no irrigation in her field, she brought water in by donkey, digging holes one by one and carefully filling them with seeds. Some plants grew, others withered and died. But turning this handful of seeds into a livelihood wasn't the women's only battle because their husbands weren't exactly cheering them on either, grumbling that they shouldn't have given up their day jobs so hastily. Still, they persevered.

Just as their efforts were taking root, literally and figuratively, the Ministry of Agriculture arrived with orders to destroy their entire crop. There was a law, they explained, prohibiting the cultivation of native cotton. The government sees cotton's promise, but their field of vision is confined to the commercial pima variety, which only grows white. Those native colours might cross-contaminate, muddying its coveted whiteness, shortening the extra-long staple fibres or harbouring pests that could damage the lucrative crop. And they might not, Yolanda thought bitterly as months of backbreaking work went up in smoke. Well, perhaps not quite all, because each cotton boll produces numerous seeds and it wasn't hard to find a few escaped handfuls. This native cotton, this country cotton, this cotton the government disparages is proving to be a plant so tough it will withstand all attempts to eliminate it. Yolanda's anger solidified into determination. She would not let them extinguish biodiversity and cultural heritage in a single swipe. If the law required her to grow white cotton, then that's what she would do. She planted a field of pure white pima cotton, just like the government wanted. And in the middle, she grew her country cotton, her coloured cotton, her cultural heritage, wrapped like a secret within a veneer of conformity.

The women tended their crops, harvesting, spinning, storing, anticipating the day the world would catch up. After eight years they grew tired of waiting and invited the technical coordinator from the Ministry to come and inspect their plants. To the government's surprise, these women had been right all along: native coloured cotton posed no threat to the white variety after all. In truth, it had some distinct advantages. While its staple length is short, native cotton is naturally resistant to the pests that plague commercial varieties. It thrives with minimal maintenance, requiring no fertilisers or pesticides, and the bushes can be harvested for up to six years. Unlike the monoculture of pima, native cotton is often planted as hedgerows to protect other crops from foraging animals. The government reversed the ban. Then they went further, officially recognising native cotton as the genetic, ethnic and cultural heritage of Peru. Yolanda and the other women had done more than rescue five colours from the verge of extinction; they had woven their efforts into history, reclaiming their cotton as a national treasure.

---

As Yolanda tells me her story, she sits on a woven petate palm mat, a strap around her back suspending her loom between her body and a nearby tree. Her shuttle moves to and fro, weaving alternating stripes of beige, brown and white. A young boy sits in the shade of her loom, while a girl sidles up to lean on her grandmother's shoulder, absorbing the craft. Not that this is surprising – they've been surrounded by coloured cotton since the day they were born. In their culture, this fibre is for more than weaving. When a baby is born, the father will bury the placenta while the mother bathes herself and her baby with purple cotton. Then she wraps a pillow or hat made from brown cotton around the newborn's head – no other colour will do. This cushions the baby's soft skull against the nocturnal hooting of the great horned owl, which could easily split it in two. As the child grows, a loop of cotton around the ankle will ease leg pain, while a wad of cotton soaked in alcohol soothes toothache and mouth ulcers, always burned after use. A compress for bites, a remedy for fright, it seems there's nothing native cotton cannot do.

Today the women's group that met on the roadside every Tuesday has grown into a movement, encompassing not just her community but twenty others besides. With the money she earns, Yolanda has sent one of her daughters to medical school, while another works beside her growing cotton. As Yolanda pushes her bobbin through the warp threads, pulling the batten towards her to tighten a strip of rich chestnut weft, her heritage comes alive on the loom. Sometimes she thinks back to life in the chicken factory. But not very often.

## An instrument of peace

***El Salvador • 2024***

El Salvador breathes as uneasily as its volcanoes. More than three decades have passed since peace accords were signed between the government and guerillas, ending a turbulent thirteen years that bled the land dry. Well, perhaps not quite dry, because the rivers still ran – red with the blood of the silenced. Seventy-five thousand lives were lost; countless more vanished into the death squad void. The bridge to peace is rarely an easy crossing, especially with wounds as deep as the country's volcanic valleys. After the war, it wasn't only communities that were divided; the economy had been devastated, the fabric of the nation torn apart. For years, violence lingered like the ashes of a volcanic eruption. When the US started deporting war refugees with criminal records, gangs from Los Angeles took root in El Salvador and homicide rates spiked to the highest in the world. Even though a crackdown on gang violence had brought calm to the streets, El Salvador's past and present struggles remained tightly entwined.

I had travelled there to meet Rhina Rehmann, a remarkable woman whose vision is about as vast as the jungle that surrounds her town and whose resolve is as firm as the volcanic rock beneath her feet. Rhina is proprietor of one of only two farms worldwide producing certified organic indigo. Located in Suchitoto, a sleepy colonial town bordering a vast lake, Los Nacimientos hacienda stands as a symbol of renewal. As we strolled past rows of indigo plants, Rhina began to share her story.

---

The war didn't just take the land; it took everything. The guerrillas drove Rhina's family from their farm and moved in, turning Hacienda Los Naranjos into an encampment. As clashes between the guerrillas and army intensified, farm workers abandoned their fields, too afraid to return. And without work, there was no food. Faced with an impossible choice, Rhina's father struck a deal, one that would hold for twelve long years: the guerrillas would allow the workers safe passage to cultivate the land, so long as the family never set foot on it again.

By the time the Peace Accords were signed in 1992, Rhina had built a life for herself in Germany, working at the Salvadoran embassy. It was there, far from home, that her destiny called her. The German government approached her with a proposal to develop an indigo project aimed at reintegrating former soldiers and guerrillas into civilian life. 'They gave me a night to think about it,' she recalled with a wry smile. 'I didn't tell my parents. They would have died of fear thinking I was working with the guerrillas.'

The initiative proved a success, but the official mission ended and with it the indigo pilot. Yet Rhina had seen indigo's potential, not only as a pigment but as a pathway out of poverty. She knew what she had to do. Although her family's hacienda had been razed to the ground when the guerrillas abandoned the property, the land remained and that was what mattered. She renamed the property Los Nacimientos, The Dawnings. Here she would cultivate not just indigo but a reawakening.

And that's how it started, like a wavering thread. Not that the indigo was any good at first. Years of intensive sugar cane farming had depleted the soil. The first harvests were poor, the pigment weak. Hope stirred when a German company reached out, desperate for a reliable supplier after others had adulterated their indigo bricks with mud to increase the weight. Rhina shipped her samples. The feedback was blunt: terrible quality, only twenty-three per cent indigotin, the blue pigment concealed in the plant's leaves. Rhina didn't even know how to measure quality back then, but their critique was a call to action; she set off in search of lost expertise.

Rhina's quest to produce the finest indigo took her deep into El Salvador: to the rugged hills of San Vincente where indigo thrived; to Zacatecoluca and Morazán, the former heart of the guerrilla communication system; to Chalatenango, where indigo and corn were interwoven. With all this wisdom wrapped in her mind, she returned to Suchitoto, where she gathered the community to plant row by row – one of maize, another of indigo – just as she had seen in Chalatenango. The following year they rotated the two crops, and, without any fertilisers, the maize grew a foot taller. The old ways, she realised, were wiser than anyone knew. Yet local farmers were slow to trust Rhina's vision, their memories of indigo still enmeshed with a history of exploitation. 'Get it out of your head and into your hands!' she urged them. Little by little, their knowledge emerged: they began to open up about how their parents and grandparents had worked with the potent blue herb. As their fear faded, the quality improved. And blue returned, rich and true. The German company placed its first order.

One barrier still remained: convention. 'Have you ever let them try?' Rhina challenged the men as they scowled at the idea of women working alongside them. Of course, Rhina knew the popular saying in her country, *either all in the bed or all in the soil*, but now was time to rewrite the rules. If women worked the land, men might show them a little more respect. Until that time, only men had beaten the indigo tanks, churning the sea-green liquid with great paddles until it turned blue. This is where Rhina's revolution would start: she insisted at least one woman should take up the paddle.

Over the years, Los Nacimientos became more than a farm, and indigo became more than an economic endeavour. As former soldiers and guerrillas worked side by side, the fractured community was stitched back together. The soil, strengthened by regenerative practices, began to recover as well, and by adopting organic cultivation, the hacienda gained a competitive edge over large producers in India, China and Japan. Luck was on El Salvador's side this time: the revitalisation of the industry coincided with a surge in demand for high-quality indigo. Those who appreciated fine denim wanted the real thing, not the synthetic version with benzene, formaldehyde and cyanide.[2] The price of indigo soared, outstripping even coffee. And so, Los Nacimientos lived up to its name: a dawning. Streams of blue gold flowed once again, carrying the promise of a future reclaimed.

---

From El Salvador, I travelled north to Guatemala, another Central American nation shaped by a long civil war and the heavy hand of US intervention. These conflicts didn't just take lives: they disrupted ecosystems, dismembered cultures and cut short the threads of ancestral knowledge, leaving a legacy of displacement that far outlasted the fighting. But traces endured, as Rhina discovered, both in the memory of the people and in the materials they worked with, drawing the past into the present.

## What existed before can be reborn

***Guatemala • 2024***

This guiding philosophy animates the work of Olga Reiche. I had admired her work for decades, even writing about her in my masters dissertation back in 1990. But nothing could have prepared me for the fascinating world to which she introduced me, one where corn silk, dental floss and leftover tortillas become materials for artistic reinvention.

---

For thirty-five years, Olga has been working to resurrect the art of natural dyeing, breathing vibrant life back into a tradition that nearly vanished from Guatemala. Many believed natural colours could never be resuscitated; some say this still, and not without reason. Olga herself admits that despite her tireless efforts, natural dye practice has been a failure, even in groups where her training was initially well received. Leaders stray from environmentally friendly methods, workers fail to follow instructions and market stalls still overflow with synthetic fibres and AZO dyes, synthetic colourants with significant environmental and health risks. 'And Chichicastenango?' I ask, my mind flashing back to the colourful photographs I took three decades ago. Olga's response saddens me: the largest market in Central America has lost much of its charm, now cluttered with a lot of junk. Yet Olga perseveres, driven by an unswerving passion for hand-weaving and natural dyes.

Olga is no stranger to innovation; it's what she's been doing ever since she opened her shop Indigo in Antigua in 1987 to support the widows of Guatemala's civil war. But the fibres on her loom when I visit are something entirely different, pushing the boundaries of craft even further. This time, she isn't just rediscovering ancient techniques; she is transforming the unexpected. I have to say I never imagined dental floss would find a place in this book, but of course, this is Olga Reiche, experimenter supreme. It all started, as many of Olga's stories do, with a chance encounter. A woman giving up her eco-business offered Olga hundreds of small pots of corn-based dental floss. Many would consider this a thread too far, no matter how many *chicharrones* you eat, but Olga saw these pots as treasure troves of potential. Twelve of those spools have already been used by Sabina, a young artisan from the community, who sits working on a backstrap loom. Olga enthuses about the fibre's possibilities. 'Look!' she says, indicating a curl with a soft coral sheen. 'It dyes beautifully.' She's right. Laurel and cochineal coax radiant pinks, fustic produces a golden yellow, while *palo de vida* is glossy as a conker newly released from its spiky shell.

But corn's role in textile production doesn't end there. The weavers of Antigua often work with single-ply cotton, a fragile fibre prone to breaking on the loom. Here, families consume countless tortillas, some of which invariably remain at the end of the day. By morning they are crisp and dry – many would simply throw them away. Not Olga. Through experimentation she discovered that soaking old tortillas in water produces a gummy liquid. When you add cotton threads to this mixture, they grow strong and supple, easier to weave. As if these aren't sufficient examples of corn's transformative power, Olga found silk drawn from corn husks produces a dye.

---

I learnt so much from Olga, and not just about corn. We delved into the use of tannin-rich plants in the dye process, including the trunk of the banana plant, which is cut down as soon as the fruit are harvested. Use it quickly and it fixes the colour without altering the shade; leave it too long, as most people do, and the fibres take on a brownish hue. Naturally, we also discussed indigo, Olga's favourite pigment and the name of her shop, which is filled with the most exquisite weavings, as you'd expect.

Olga's work embodies a belief in second chances, not only for traditions and materials but for communities and ecosystems. Every discarded object, every overlooked material, holds the potential for revival. In her hands, even the mundane becomes magical. What existed before can indeed be reborn, not as a shadow of its former self, but as something entirely new and full of possibility.

## I weave for a better future

*Mexico • 2024*

How do we think and create like artists? Angela Damman challenged a group of young Maya women preparing for their first major exhibition. The question, much like the henequen fibre they work with, carried layers of meaning. How do we take what is humble and raise it to the level of art? How do we demystify a craft long confined to the margins and make it accessible and aspirational for a new generation? For one of the apprentices, nineteen-year-old Wendy Guadalupe Dzul Can, this question was deeply personal. She had joined the Maya Youth Artisanship Initiative five years earlier, drawn by a desire to learn the weaving her grandmother practised. Now, about to become a mother herself, she embraced the challenge with both hands.

The roots of this story reach back further, to a serendipitous encounter on a village road. One day, Angela spotted a man carrying tools in a woven *sabucán* and stopped him to ask what his bag was made from. When she learnt it was made from local henequen, Angela, a sustainable designer with farming roots, recognised untapped possibilities for this near-forgotten plant. 'Artisans are the guardians of one of the Yucatán's most valuable resources,' she told me as we sat in her studio, sansevieria fibre chandeliers floating like comets overhead. The arrival of synthetic fibres devastated the henequen industry and, as cultivation collapsed, the art of spinning and weaving the fibre almost vanished. By setting up the Maya Youth Artisanship Initiative, Angela is working to reverse that decline, convinced the plant has potential beyond our imagining. More than sackcloth and string, it can be anything.

For young Maya women like Wendy, the obstacles are many: Indigenous, poor, rural and speaking Maya, they face exclusion at every turn. This initiative seeks to turn these perceived disadvantages into strengths, converting rural roots into pathways of opportunity. Wendy's journey reflects this transformation. 'I weave for a better future,' she says, smiling. 'My grandmother inspired me to weave.' Her grandmother, María Reyes Maas Cab, a mentor in the programme, is proud of Wendy's progress. 'I worked well with my granddaughter, and she also worked well with me. I liked working as a team like that.'[3] Across the generations, these women began weaving not just a textile, but a legacy.

As the young Maya women became more confident in their weaving and natural dye skills, Angela saw they were ready. They would put on an exhibition at the Museo de Arte Popular in

Mérida called *Weaving Who We Are.* When she posed the question about how to think and create like artists, the women rose to the challenge. Botanical dyes gave expression to their work, gathered from their gardens, the cornfield, the jungle: *chukum* for beige, achiote for orange, *tsalam* for smoky reds, *cholul* for soft pink, *muicle* for purple, *chak te* for bronze and *sabacché* when the fibres are gold-spun. Whatever colour they conjure up in their minds, nature meets their creativity with a shade to match.

When the exhibition opened, Wendy's was the first work to sell, purchased by an American editor known for high-end architectural magazines. Soon, every piece had found a home. The women were overjoyed, but this was about more than the economic opportunity. As the exhibition's title suggests, for Wendy and her peers, henequen is more than a thread to the past; it stretches far beyond the loom, connecting past to future, tradition to empowerment, earth to art.

'I wish for this to never end,' said Wendy, now fully committed to her craft. As she prepares to one day teach her child, she has become more than an apprentice weaver; she is a guardian of roots and originator of new shoots.

---

While staying with Angela, I also wanted to find out about the broader ecosystem supporting her work, because henequen is part of a bigger picture of agriculture, science and innovation. To understand more, I travelled to the Yucatán Scientific Research Centre in Mérida where the plant is undergoing a transformation of its own.

'This agave is not a clone,' Dr Lorenzo Teyer explained, cradling a laboratory-grown seedling. 'It's a container for genetic diversity.' Within its plastic protection, the green shoot, no larger than a finger, seemed unremarkable. But for Dr Teyer, it was no less than a revolution. Each seedling belongs to a new generation of henequen plants, carefully adapted to modern agroecological needs. Thousands of these seedlings filled the artificially lit shelves – it's about as far from my visit to the sunlit Maya village as you can get, yet equally important. 'We've achieved remarkable results,' Dr Teyer continued, describing how they already had some two million plants in the countryside, including espadín agave used by the mezcal industry. 'Thirty per cent more leaves a year, a twenty-five per cent increase in leaf length in the first two years and a harvesting cycle that begins in just two years instead of four. This is groundbreaking.' These advances mean producers now get twice the amount of fibre. It's little wonder around two-thirds of henequen in Yucatán now comes from micropropagation. But the laboratory's work goes beyond immediate yield, focusing also on preserving genetic diversity.

As I walked through rows of seedlings in the experimental laboratories, each with their own birth records and data sheets, I felt the convergence of the old and the new. Here innovation was measured in millimetres and molecular changes. But the goal was the same: a vision of where plant fibres can take us. Every leaf, every fibre, holds latent potential. In Dr Teyer's laboratory, henequen becomes invention. In Angela's vision, it becomes art. In Wendy's hands, it becomes identity. Henequen is not bound to the past; nor is it constrained to the present. It is a fibre of reinvention, endlessly rewoven.

# CONCLUSION
# Last Strands

The year has turned. Outside my window, the snow falls in great, weightless flakes. It has been coming down for hours. Droplets of moonlight bleed through the ghostly trees, their branches dressed in winter-white mufflements, and splash onto the powdery quilt beneath. I turn on an outside light to let the dog out. Earlier, dark stalks of flattened angelica and moor grass served as a constant reminder to pause my writing and tend the chaos in my garden. Now the world is unbroken whiteness, as if drawn anew. It hides not just the remnants of autumn's neglect, a consequence of living in my caravan, but also the scrabbling of small creatures finding food to survive, an understory of invisible beings that transform the way things work. This pure stillness is an illusion, blanketing my garden's dishevelment. And it is an invitation to look beyond appearances.

I think back to my reflection on snow cloth in Echigo, Japan – a story that made me question the true nature of colour – and I suddenly realise this is no white garden. Each snowflake is a fleeting prism, its ice crystals scattering light to the full rainbowed spectrum. The colours are reflected so many times that they all blend together, returning to my eyes as the perception of white. Snow is a paradox of whiteness, revealing and concealing something at once simple yet infinitely complex. What appears pure and pale by moonlight contains every hue in the spectrum. As the villagers of Echigo know full well, snow shimmers, refracts and dances with light, a shifting presence that transforms snow cloth into a reflection of itself.

In the same way, the threads that connect us with plants are rarely simple or singular; they are spun and plied, woven into patterns that defy categorisation, transcending family, genus or kingdom. What appears to be one story often contains multitudes, just as snow is never simply white. To look closely at a single thread, as I did when I peered down the microscope at that twist of blue cotton, is to uncover connections that span space and time. A thread might appear frail yet be vital to the integrity of the weave. Or it might seem ephemeral, only to endure, persisting within the environment or within the fabric of culture. If it happens to be the first length of string ever spun, it may prove to be fundamental to our existence on this Earth. A single thread and a snowflake – both hold more than they seem.

Imagine a centuries-old textile, like an antique tapestry. Now turn it over in your mind and explore the reverse, running your fingers across the threads, seeing how it was made, and remade. On the surface, it may look pristine, like the snow in my garden, but beneath lies a tangle of loose threads, knots, centuries of care and repair. Each strand tells a story; remove one and the whole fabric might unravel. In the same way, our relationship with plants is not a linear narrative of progress but a tapestry of contrasts, where light and shadows intertwine. It is a story of mistakes and learning, frailty and resilience, strands cut, left hanging or repaired. Sometimes a strand even doubles back on itself and heads off the wrong way.

This journey began with a question: what if the story of humanity could be told through a single thread? That thread has carried us across centuries, from the earliest evidence of Neanderthal fibre technology to biomaterials and regenerative textiles. We have travelled continents and oceans, from the Amazon basin to the

Great Smoky Mountains, from a sandstone escarpment in Mali to the royal palaces and odoriferous backstreets of France, from the windswept bays of New Zealand to snowbound villages in Japan. Everywhere we have seen how fibres and dyes are more than commodities. When land is stripped bare and sapped dry, when profit eclipses people, they leave indelible scars. But we've also seen how such materials can restore what has been lost, regenerating ecosystems, enriching soils and strengthening communities. Like snowflakes scattering light into the full spectrum of colour, these stories resist simplification, urging us to look more deeply and question what we think we know

For too long, we have framed nature as separate from humanity, a resource to be controlled or consumed, perpetuating a cycle of extraction and exploitation. But other ways of seeing are possible. The Guaraní understand existence as a process of becoming rather than being, a continuously unfolding relationship with the natural world. The fibres of their culture have been severed, but still the people hold strong, refusing to break. Their fight today is about more than reclaiming ancestral lands; it's about restoring integrity to a woven way of life. It is about retracing their steps to the Land Without Evil, where the past and future intertwine, trusting the shiny red dye seeds to show them the way. Their philosophy challenges Western obsessions with permanence and stability, our seeking of unchanging truths and resistance to the inevitability of change. It demands a shift in perspective, like the inside-out reading of Guamán Poma's *mappa mundi* in his chronicle of the world turned upside down. Can a single thread tell our story? In the end, perhaps it all comes down to the way we see. We can read this book as a history of plant materials. Or we can see it as a map for renewal, revealing myriad pathways towards a more connected future.

I think back to the original meaning of the word map as a fluttering flag, a streaming cloth unfurling through time, revealing the patterns of our passage through the world. Lines on a map give form to the ground beneath, yet the shifting soils beneath my Undercliff caravan remind me that nothing is truly still, nor ever fully known. Earth moves, paths cross, stories interweave, altering the landscape for those who will follow. Like a banner raised to the wind, I hope this book does more than recount what has been and acts as a guide of sorts to what might yet be.

If told through a single thread, the narrative of humanity is not a linear tale. It twists and twines into a kaleidoscope of patterns and dazzling contrasts. It is not static but alive, interwoven with the choices of the past, the challenges of the present and the hopes of a future still to be written. It undoubtedly consists of multiple strands plied together – it may even be white – but it urges us to acknowledge the full spectrum of colour as a dynamic interplay that holds space for hope and complexity. Perhaps, if we let it, this thread is strong enough to bind us and the plant world back together, refashioning the concept of nature as in earlier definitions, indivisible from humankind. Born of soil and seed, this thread carries within it the potential to transform not just our landscapes, but the way we see ourselves and our place in the world. It is an invitation: to trace, to weave, to create.

As the snow continues to fall, I no longer see a white garden. I see a textile of complexity and infinite possibility. Snow is not white. A single thread does not run straight. Both encourage us to look deeper, to uncover the full spectrum of life's hues. To find beauty in the interwoven, multicoloured fabric of life, firmly rooted in wonder.

# ACKNOWLEDGEMENTS

Where does a book begin? In many ways *The Nature of Fashion* has its roots in my masters in Native American studies at the University of Essex, where Valerie Fraser gave me the freedom to explore a range of themes, including the rich world of Andean textiles and natural dyes. She taught me Quechua as well. As I was the only student on the programme – the study of Indigenous cultures being less fashionable in those days – I joined the PhD tutorials of the late Professor Gordon Brotherston, in which he unravelled the complexities of the codices. Many of these stories trace their origins to that formative time.

Years later, garden designer Lottie Delamain rekindled my passion for plants during our collaboration on A Textile Garden for Fashion Revolution at RHS Chelsea Flower Show. Kate Turnbull of The Secret Dyery joined us, and as we immersed ourselves in a world of botanical colour, it deepened a fascination that had never truly faded. In its earliest iteration, this book was a plant compendium, a collaboration between the three of us. I am forever grateful to Lottie and Kate for taking those first steps alongside me before encouraging me to branch out on my own. As I made my own way, I drew inspiration from the work of Uruguayan author Eduardo Galeano, whose *Memory of Fire* trilogy tells the history of Latin America through a mosaic of stories.

My curiosity about microfibres was grounded in science thanks to Emily Penn and Lucy Gilliam, who selected me to take part in eXXpedition and helped me reconnect with the ocean. Later, I returned to the water with Professor Deirdre McKay (Keele

University), Dr Antonia Law (University of Nottingham) and Dr Tom Stanton (Loughborough University) during Fashion Revolution's Restorying Riverscapes project. Our ongoing study of natural fibres in the sediment continues to inform my work today.

I want to extend my sincere thanks to the Churchill Fellowship team, whose enthusiasm for my exploration of alternative fibres and dyes, and generous support for my travels enriched this book. The knowledge I gained during that pivotal period interweaves every chapter. I remain deeply grateful to the artists, weavers, dyers, entrepreneurs, scientists and students who so generously shared their expertise. In Peru, special thanks go to Yeraldin Morillo, Carola Solis, Karin Santa Maria and Cusi Saco Chung of Mosqoy Peruvian Textiles for their support with my itinerary, transport and hospitality; to Sadith and Olinda Silvano of the Shipibo-Konibo community of Cantagallo, Lima; to Vabi Miguel Toribio and Eva Ortiz Rigle of the Yanesha Community of Santa Rosa de Pichanaz, Santo Domingo; to Yolanda Contreras from Mórrope, Lambayeque and Círculo de Investigación en Algodonero de la Facultad de Agronomía at Universidad Nacional Agraria La Molina, Lima, for expanding my knowledge of Peru's coloured cotton. My thanks also go to Rhina Rehmann and Grace Guirola in El Salvador, and to Olga Reiche in Guatemala for teaching me about indigo and beyond. In Mexico, I am particularly grateful to Alfredo Orozco of Orozco Textil Experimental, Vicente Reyes and Verónica Valdés of Fibra x Hermano Maguey, and the staff at Centro de Investigación Científica de Yucatán, Mérida. Special thanks to Angela and Scott Damman for ongoing friendship, hospitality and for passing on their passion for henequen. Also, to the Fashion Revolution teams around the world who hosted me over the years, from Bangladesh to the Philippines. Special thanks to Jesica Pullo who procured nearly every book available on natural dyes in Argentina.

# *Acknowledgements*

Some of these stories were germinated in the fanzine *Fashion Craft Revolution*, especially the section on Intangible Cultural Heritage, and I extend my thanks to the Fashion Revolution team for entrusting me with that publication. I am also grateful to Rebecca Harfield and everyone at the Royal Botanic Gardens Kew for revealing some of the treasures of the Economic Botany collection and creating opportunities for new collaborations.

*The Nature of Fashion* would not have reached publication without my amazing literary agent, Julia Silk, whose love of fashion and unwavering belief in this book carried it forwards – and me along with it. Thank you, Julia. And to Muna Reyal, my editor at Chelsea Green, for loving my writing and gathering up all the loose strands into a cohesive whole. Also, to Susan Pegg for your thoughtful editing of the text, and all the fabulous team at Chelsea Green who have worked to publish and promote this book. And to Victoria Villasana for the beautiful artwork that captures the essence of my book. I am so grateful.

Closer to home, there are so many friends and family who have supported me. Judy Scott-Moncrieff, queen of all hearts, thank you for love and laughter. To her Friday coffee girls, with Humphrey and Geoffrey, wise among us, for keeping life in perspective and ushering in the weekends with a smile. And to Maria for cake.

To Branscombe, where home has been my caravan for eight months a year, I offer thanks to those who tended this patch of earth before me and to everyone at the Sea Shanty for helping me realise this long-held dream. Without Undercliff, my caravan, a place where time is marked by the tides and the only interruption is the peregrine's cry, this book may never have seen the light of day. I owe much to my parents, who set me walking down this village's donkey paths before I could ask where they led; to my father, for teaching me the names of butterflies and wildflowers; and my mother, who instilled in me the belief that I could do anything. To my Branscombe forebears, who revealed the hidden world of these cliffs and showed me that the humble winkle was food fit for a queen. And to Mother Nature, the best teacher of all.

To Mark – husband and chef extraordinaire, forager, fisherman, winklepicker, lobster-catcher, limpet-priser – whose feasts never fail to delight. Thank you for your love, for supporting my uprooting to Branscombe and for giving me space to write. For letting me take Romero, our bearded collie, ensuring I spent almost as much time walking the cliffs and beach as I did with a book or laptop. For journeying with me to Guatemala and Mexico, navigating remote villages in search of fibres and dyes, never questioning my obsession, only asking: where next?

To my daughter, Sienna, whose passionate campaigning for nature is matched by her love of the moors and the sea. Always drawn to the waves, she reminds me that our sense of belonging isn't anchored to a fixed place but flows like the tide that etches its own stories onto Branscombe's shifting shore. With her partner, Ross Miller, and my stepson, Jade Forest, his wife, Joanisabel, and their son – our first grandson – Ambrose Silver, you remind me that this story is endlessly unfolding.

# GLOSSARY OF PLANT NAMES

Translations of plant names are based on the plant's geographical context within the book.

**ABACA** *Musa textilis:* species of banana, which produces Manila hemp. Sap from *Musa* spp. is used as a mordant and dye.

**ACHIOTE** *Bixa orellana:* annatto, lipstick tree, uru'ku, urucum, yetsap, guandur, huantura, mandur.

**AGAVE** *Agave* spp. (esp. *A. americana*): century plant, maguey, metl, pita, ixtle.

**AGUAJE** *Mauritia flexuosa:* ité palm, ita, buriti, muriti, miriti, canangucho, moriché, acho.

**ALKANET** *Alkanna tinctoria:* dyer's alkanet, dyer's bugloss, orchanet.

**AYRAMPU** *Tunilla soehrensii:* airampo cactus.

**BLACK WALNUT** *Juglans nigra:* Eastern black walnut, se-di, nogal negro.

**BLOODROOT** *Sanguinaria canadensis:* bloodwort, red puccoon root, tetterwort, dragons blood, redroot, gitli uwatali.

**BRAZILWOOD** *Paubrasilia echinata:* pernambuco, pau brasil.

**CHAPI** *Rubiaceae Galium* or *Relbunium microphyllum:* chapi-chapi, chamiri, antanco. A close relative of madder.

**CHICA** *Fridericia chica:* cricket vine, puca panga, carajuru.

**CH'ILLCA** *Baccharis latifolia:* arbusto bajo, chilca.

**COTTON** *Gossypium hirsutum:* upland cotton, Mexican cotton, algodón, comprising over ninety per cent of global cotton production.

**COTTON (NATIVE)** *Gossypium barbadense:* algodón nativo, algodón El País, bes.

**FLAX** *Linum usitatissimum:* flaxseed, linseed.

**GENIPAP** *Genipa americana:* jenipa, jagua, huito, huitol.

**HEMP** *Cannabis sativa:* má, dàmá. Also *Apocynum cannabinum* or Indian hemp, dogbane, weihkippeis.

**HENEQUEN** *Agave fourcroydes:* sak ki, sisal (named after its export port); note that *Agave sisalana* is a different species.

**INDIGO** *Indigofera* (esp. *I. tinctoria* L. and *I. suffruticosa* Mill.): añil, xiquilite, jiquilite, ch'oh, neel.

**JUTE** *Corchorus olitorius:* Tossa jute, jute mallow. Also *C. capsularis*, white jute, huángmá.

**KAPOK** *Ceiba pentandra:* ceiba, tunsho, silk cotton tree, huimba, lupuna.

**KCULLI** *Zea mays:* maíz morado, purple corn, olote.

**KINSA K'UCHU** *Baccharis genistelloides:* carqueja dolce, carqueja amarga, charara, carqueja. The dye is produced by a fungus that grows on the leaves.

**LOGWOOD** *Haematoxylon campechianum:* bloodwood, palo de Campeche, palo de tinto, ek.

**LOTUS** *Nelumbo nucifera:* padonma kya.

**MADDER** *Rubia tinctorum:* dyer's madder, garance des teinturiers. Also wild madder (*R. peregrina*).

**MOLLE** *Schinus molle:* Peruvian peppertree, huiña, muli, cullash.

**MOTE MOTE** *Vaccinium floribundum* Kunth: Andean blueberry, mortiño.

**NETTLE** *Urtica* spp.: European nettle, stinging nettle (*U. dioica*); ramie, zhùmá (*Boehmeria nivea*); Himalayan nettle, Nilgiri nettle (*Girardinia diversifolia*).

**NEW ZEALAND FLAX** *Phormium tenax:* harakeke, naut. New Zealand hemp.

**ÑUÑUNQA** *Solanum nitidum*: ñuñumaya.

**PANAMA HAT PALM** *Carludovica palmata:* paja toquilla, bonbonaje, jipijapa.

**PAPER MULBERRY** *Broussonetia papyrifera:* used for making barkcloth, known as nioge or tapa. Other species of rainforest fig may also be used.

**Q'AQA SUNKA** *Usnea barbata* and *Tillandsia capillaris:* rock beard, Spanish moss, barba de árbol, barba del cerro, barba de la roca.

**Q'OLLE** *Buddleja coriacea:* k'olle, colle, matico.

**RAFFIA** *Raphia* spp.: various species, including *R. vinifera* and *R. farinifera,* are used to make Kuba cloth.

**RUBBER** *Hevea brasilensis:* seringueira. Produces latex, which is also known as Indian rubber, caucho, caoutchouc, shiringa.

**SAFFRON** *Crocus sativus:* saffron crocus, za'faran.

**SANSEVIERIA** *Sansevieria trifasciata:* snake plant, mother-in-law's tongue, lengua de vaca.

**SPANISH BROOM** *Spartium junceum:* weaver's broom, rush broom, genêt d'Espagne, retama.

**TARA** *Caesalpinia spinosa:* peruvian carob, spiny holdback.

**THIRI** *Miconia andina:* thiri thiri.

**WELD** *Reseda luteola:* dyer's rocket, dyer's weed, yellow weed, buí mór.

**WOAD** *Isatis tinctoria:* dyer's woad, dyer's weed, glastum, pastel des teinturiers.

**YANALI** *Bocconia frutescens:* parrotweed, plume poppy, tree poppy, chilacolote blanco, pan cimarrón.

# NOTES

## INTRODUCTION: A SINGLE THREAD

1. Allen J. Christenson, trans., *Popol Vuh: Sacred Book of the Quiché Maya People*. Adapted from translation notes by Christenson (University of Oklahoma Press, 2003), 50, http://www.mesoweb.com/publications/Christenson/PopolVuh.pdf.
2. Allen J. Christenson, email message to author, 18 December 2024.

## PART ONE: IN THE BEGINNING WAS THE CLOTH

1. B.L. Hardy, et al., 'Direct Evidence of Neanderthal Fibre Technology and Its Cognitive and Behavioral Implications', *Scientific Reports* 10, 4889 (2020), https://doi.org/10.1038/s41598-020-61839-w.
2. Elizabeth Wayland Barber, *Women's Work: The First 20,000 Years: Women, Cloth, and Society in Early Times* (W.W. Norton & Company, 1994), 42.
3. William Morris, 'The Beauty of Life', lecture, Birmingham Society of Arts and School of Design, 19 February 1880, in *The Collected Works of William Morris: Hopes and Fears for Art. Lectures on Art and Industry*, ed. May Morris (Longmans Green and Company, 1910), 53.
4. Tiago Barros Afonso, et al., 'Phenolic Compounds from By-Products for Functional Textiles', *Materials* 16, no. 22 (2023): 7248, https://doi.org/10.3390/ma16227248.
5. Jennifer Hattam, 'What Happened to Turkey's Ancient Utopia?', *Discover*, 28 July 2016, https://www.discovermagazine.com/the-sciences/what-happened-to-turkeys-ancient-utopia.
6. Lambert, A.F. 'A Possible Place-Sign (Toponym) from Oxtotitlán Cave, Guerrero, Mexico', *The Post Hole*, no. 27 (February 2013).
7. Thomas Campbell, *Letters from the South*, vol. 1 (first published in *New Monthly*, 1835), 150.
8. Francis Grose, *A Classical Dictionary of the Vulgar Tongue* (S. Hooper, 1785), 141.
9. Christian Bergfjord, et al., 'Nettle as a Distinct Bronze Age Textile Plant', *Scientific Reports* 2 (2012), https://doi.org/10.1038/srep00664.
10. Albert Einstein, 'What Life Means to Einstein', interview by George Sylvester Vierick, *The Saturday Evening Post*, 26 October 1929 (Philadelphia).

## PART TWO: AND COLOUR FLOWED THROUGH EVERY LAND

1. Peter Gouras in *Color Vision*, Webvision.med.utah.edu.
2. Barry Donovan, 'Chomh Buí leis an mBuí Mór', *The Irish Echo*, 16 February 2011.
3. G.K. Chesterton, *Tremendous Trifles,* 10th ed. (Methuen & Co., 1927), 5–6.

## PART THREE: WEAVING SHADOWS AND LIGHT

1. 2 Esdras 6:42 Authorised Version (Apoc).
2. Isaiah 60:5 AV.
3. Bartolomé de las Casas, *The Diario of Christopher Columbus's First Voyage to America*, 1492–1493, trans. Oliver Dunn and James E. Kelley Jr. (University of Oklahoma Press, 1989), 65.
4. Amerigo Vespucci, 'The Medici Letter: Letter on His Third Voyage to Lorenzo Pietro Francesco di Medici', March (or April) 1503, in *The Letters of Amerigo Vespucci*, ed. and trans. Clements R. Markham (Hakluyt Society, 1894), Early Americas Digital Archive, 2005.
5. Bartolomé de las Casas, *Brevísima Relacion de la Destrucción de las Indias*, trans. Francis Augustus MacNutt (Project Gutenberg, 1909), https://www.gutenberg.org/cache/epub/20321/pg20321-images.html.
6. Bartolomé de las Casas, *Obras Completas*, vol. 1, *Vida y Obras*, ed. Álvaro Huerga (Alianza Editorial, 1998), 383.
7. Her Majesty's Stationery Office, *Letters and Papers, Foreign and Domestic, Henry VIII*, vol. 10, January–June 1536 (London, 1887), 313, https://archive.org/details/letterspapersfor10greauoft/page/312.
8. W.S.W. Vaux, 'Some Notices of Records Preserved Amongst the Corporation Archives at Southampton', *The Archaeological Journal* 3 (1846): 231, https://doi.org/10.5284/1066809.
9. 'By the Queene. A Proclamation Inhibiting the Sowing of Woad', *Early English Books Online 2* (University of Michigan Library Digital Collections, [1600]), https://name.umdl.umich.edu/A21946.0001.001.
10. Richard Hakluyt, 'Remembrances for Master S.', 1582, in *The Principal Navigations, Voyages, Traffiques and Discoveries of the English Nation*, vol. 5 (Project Gutenberg, 2004), https://www.gutenberg.org/ebooks/7900.
11. 'Proceedings', in *Colonial Records of Virginia* (R.F. Walker, Superintendent Public Printing, 1874), 22.

## PART FOUR: EVERYTHING IS CONNECTED

1. Louis Crommelin, 'An Essay towards the Improving of the Hempen and Flaxen Manufacturer in the Kingdom of Ireland' (Dublin, 1705) in 'The French

Settlers in Ireland. No. 2. The Huguenot Colony at Lisburn (Continued)', *Ulster Journal of Archaeology* 1 (1853): 287, https://www.jstor.org/stable/20563475.

2. Harriet Horry Ravenel, *Women of Colonial and Revolutionary Times: Eliza Pinckney* (Charles Scribner's Sons, 1896), 69.
3. Harriet Horry Ravenel, *Women of Colonial and Revolutionary Times*, 38.
4. Elizabeth Donnan, 'The Slave Trade into South Carolina Before the Revolution', *The American Historical Review* 33, no. 4 (1928): 822, https://doi.org/10.2307/1838372.
5. Alexander von Humboldt and Aimé Bonpland, *Personal Narrative of Travels to the Equinoctial Regions of the New Continent, During the Years 1799–1804*, trans. Helen Maria Williams, vol. IV (Longman, Hurst, Rees, Orme, and Brown, 1819), 63.
6. Alexander von Humboldt, *Essay on the Geography of Plants*, ed. Stephen T. Jackson (University of Chicago Press, 2009), 79.
7. Alexander von Humboldt and Aimé Bonpland, *Personal Narrative of Travels*, 120.
8. Alexander von Humboldt, *Aspects of Nature, in Different Lands and Different Climates, with Scientific Elucidations*, trans. Elizabeth J.L. Sabine, vol. 2 (Longman, Brown, Green, and John Murray, 1849), 11.
9. Charles Claude Hamilton, *An Essay on the Art of Flying* (J.L. Cox and Sons, 1839).
10. Subhas Bhattacharya, 'The Indigo Revolt of Bengal', *Social Scientist* 5, no. 12 (July 1977): 14, https://doi.org/10.2307/3516809.
11. National Archives of India, *Minutes of the Local Government meetings 28 May 1860*, testimony by Panjee Mulla, National Virtual Library of India, 2.
12. Subhas Bhattacharya, 'The Indigo Revolt of Bengal', 13.
13. National Archives of India, *Minutes of the Local Government meetings 29 May 1860*, testimony by Panjee Mulla, National Virtual Library of India, 69.
14. National Archives of India, *Minutes of the Local Government meetings 28 May 1860*, testimony by Kulin Mundul, National Virtual Library of India, 2.
15. Francisco Ximénez, *Quatro Libros de la Naturaleza, y Virtudes de las Plantas y Animales* (Casa de la Viuda de Diego Lopez Daualos, 1615), 147, https://bibdigital.rjb.csic.es/records/item/13778-quatro-libros-de-la-naturaleza-y-virtudes-de-las-plantas-y-animales.
16. Margaret Atwood, 'It's Not Climate Change – It's Everything Change', Medium, 27 July 2015, https://medium.com/matter/it-s-not-climate-change-it-s-everything-change-8fd9aa671804.
17. Adam Cooper, *Green Book Fair: Eco Narratives with Adam Cooper (Threads in the Ground) and Sarah France*, online trade fair, 27 September 2023, posted 3 October 2023 by Comma Press, YouTube, 19 min., 15 sec., https://www.youtube.com/watch?v=TMz_bR4y3MU.

## PART FIVE: SHOOTS OF RESISTANCE


1. Société Nationale d'Acclimatation de France, 'Spartium', *Revue des Sciences Naturelles Appliquées*, 5 April 1891.
2. Friedrich Carl Theis, *"Khaki" on Cotton and Other Textile Material*, trans. E.C. Kayser (De Gruyter, 1903): 61–63, https://doi.org/10.1515/9783112671023-024.
3. United Nations Environment Programme, 'Critical Ecosystems: Congo Basin Peatlands', 27 February 2023, https://www.unep.org/news-and-stories/story/critical-ecosystems-congo-basin-peatlands.
4. The Proceedings of the Old Bailey, 1674–1913, https://www.oldbaileyonline.org.
5. Barbara Farquharson and Joan Doern, *The Branscombe Lace-makers* (The Branscombe Project, 2002), 2.
6. Great Britain. Children's Employment Commission (1862). *First Report of the Commissioners, with Appendix* (George Edward Eyre and William Spottiswoode, 1863), 185.
7. Geotraceability Fair Trade European Project 2009–2012, Chambre de Commerce et d'Industrie du Gers.
8. US Department of State, '416. Memorandum From the President's Assistant for National Security Affairs (Brzezinski) to President Carter', in Foreign Relations of the United States 1977–1980, Volume XV, Central America, 1977–1980, ed. Nathaniel L. Smith (United States Government Printing Office, 2016), 1052.
9. The Archbishop Romero Trust, 'The Final Homily of Archbishop Romero: Mass on the First Anniversary of the Death of Sara Meardi de Pinto', 24 March 1980.
10. María López Vigil, *Oscar Romero: Memories in Mosaic*, trans. Kathy Ogle (EPICA, 2000), 72.
11. James Brockman, *Romero. A life* (Orbis Books, 1989), 248.


## PART SIX: AS THREADS SPIRAL ON, SO DOES TOMORROW


1. Emma Hakansson, director and producer, *Shiringa: Fashion Regenerating Amazonia* (Collective Fashion Justice, 2025), online video, https://www.collectivefashionjustice.org/shiringa-film.
2. American Institute of Chemical Engineers, 'Catalyzing Commercialization: Rapid Assay Opens the Door to Natural Indigo Dye and Other Bio-Based Chemicals', *CEP Magazine*, May 2018, https://www.aiche.org/resources/publications/cep/2018/may/catalyzing-commercialization-rapid-assay-opens-door-natural-indigo-dye-and-other-bio-based-chemicals.
3. Ashley Kubley and Angela Damman, *Tejiendo Quienes Somos (Weaving Who We Are): The Maya Youth Artisan Initiative* (Museo de Arte Popular Yucatán and University of Cincinnati Office of Research, 2021).

# SELECT BIBLIOGRAPHY


Adovasio, J.M. and T.D. Dillehay. 'Perishable Technology and the Successful Peopling of South America.' *PaleoAmerica* 6, no. 3 (2020): 210–22.

Ali, T.O. *A Local History of Global Capital: Jute and Peasant Life in the Bengal Delta*. Princeton University Press, 2018.

Antúnez de Mayolo, K.K. 'Peruvian Natural Dye Plants.' *Economic Botany* 43, no. 2 (1989): 181–91.

Bergfjord, C., et al. 'Nettle as a Distinct Bronze Age Textile Plant.' *Scientific Reports* 2 (2012): 664. https://doi.org/10.1038/srep00664.

Berglund, J. 'Kirkebjerget – A Late Bronze Age Settlement at Voldtofte, South-West Funen.' *Journal of Danish Archaeology* 1, no. 1 (1982): 51–63. https://doi.org/10.1080/0108464X.1982.10589875.

Bhattacharya, S. 'The Indigo Revolt of Bengal.' *Social Scientist* 5, no. 12 (July 1977): 13–23. https://doi.org/10.2307/3516809.

Cadogan, L. 'Animal and Plant Cults in Guarani Lore.' *Revista de Antropología* 14 (1966): 105–24.

Caley, E.R. *The Leyden and Stockholm Papyri: An English Translation with Brief Notes*, edited by W.B. Jensen. University of Cincinnati, 2008. https://homepages.uc.edu/~jensenwb/books/Leyden%20&%20Stockholm%20Papyri.pdf.

Cobo, B. *Historia del Nuevo Mundo*. 4 vols. Reprint of 1653 text, D.M.J. de la Espada, Impresa de E. Rasco, 1890.

Conard, N.J. and V. Rots. 'Rope Making in the Aurignacian of Central Europe More than 35,000 Years Ago.' *Science Advances* 10, no. 5 (2024). https://doi.org/10.1126/sciadv.adh5217.

Crommelin, L. 'An Essay Towards the Improving of the Hempen and Flaxen Manufacturer in the Kingdom of Ireland. Dublin, 1705.' In C.N. de la Cherois Purdon 'The French Settlers in Ireland No. 2. The Huguenot Colony at Lisburn (continued)', *Ulster Journal of Archaeology, First Series* 1 (1853): 287. https://www.jstor.org/stable/20563475.

The Cultural Foundation for Promoting the National Costume of Japan. 'Echigo Jofu, Ojiya Chijimi.' In 'Dyeing and Weaving', *Dictionary of Kimono Terms*. https://www.kimono.or.jp/dictionary/eng/echigojoufu.html.

Dias, B. do N. 'Munduruku Cosmopolitics and the Struggle for Life.' In *Indigenous and Minority Populations: Perspectives From Scholars and Writers across the World*, edited by S.G. Barnabas. IntechOpen, 2023. https://doi.org/10.5772/intechopen.109926.

Dillehay, T.D., et al. 'A Late Pleistocene Human Presence at Huaca Prieta, Peru, and Early Pacific Coastal Adaptations.' *Quaternary Research* 77, no. 3 (2012): 418–23. https://doi.org/10.1016/j.yqres.2012.02.003.

Donkin, R.A. 'Bixa Orellana: "The Eternal Shrub".' *Anthropos* 69 (1974): 328–41.

Fagan, B. *Elusive Treasure: The Story of Early Archaeologists in the Americas*. Book Club Associates, 1977.

Folkard, R. *Plant Lore, Legends, and Lyrics: Embracing the Myths, Traditions, Superstitions, and Folk-Lore of the Plant Kingdom*. Reprint of 1884 edition, Project Gutenberg, 2014. https://www.gutenberg.org/files/44638/44638-h/44638-h.htm.

Follensbee, B.J. 'Fibrer Technology and Weaving in Formative-Period Gulf Coast Cultures.' *Ancient Mesoamerica* 19, no. 1 (Spring 2008): 87–110. https://doi.org/10.1017/S0956536108000229.

Fowler, W.R. and J.J. Card. 'Material Encounters and Indigenous Transformations in Early Colonial El Salvador'. In *Material Encounters and Indigenous Transformations in the Early Colonial Americas: Archaeological Case Studies*, edited by C.L. Hofman and F.W.M. Keehnen, 9:197–220. Brill, 2019. http://www.jstor.org/stable/10.1163/j.ctvrxk2gr.15.

Gilligan, I. *Climate, Clothing, and Agriculture in Prehistory: Linking Evidence, Causes and Effects*. Cambridge University Press, 2019.

Gilligan, I. 'The Textile Hypothesis: A Paradigm Shift for Farming Origins.' *Archaeologies* 19 (2023): 555–96. http://dx.doi.org/10.1007/s11759-023-09488-z.

Greer, A. and J. Bilinkoff, eds. *Colonial Saints: Discovering the Holy in the Americas, 1500–1800*. Routledge, 2003.

Griaule, M. *Conversations with Ogotommêli: An Introduction to Dogon Religious Ideas*. Oxford University Press, 1965.

Grömer, K. *The Art of Prehistoric Textile Making*. Veröffentlichungen der Prähistorischen Abteilung (VPA), vol. 5. Natural History Museum Vienna, 2016.

Guamán Poma de Ayala, F. *Primer Nueva Corónica y Buen Gobierno*. Reprint of 1615 edition, Siglo Veintiuno, 1980.

Haberle, S.G., et al. 'The Palaeoenvironments of Kuk Swamp from the Beginnings of Agriculture in the Highlands of Papua New Guinea.' *Quaternary International* 249, no. 6 (2011): 129–39. https://doi.org/10.1016/j.quaint.2011.07.048.

Hakluyt, R. *The Principal Navigations, Voyages, Traffiques & Discoveries of the English Nation*. Vol. 1. Ams Press Inc., 1965.

Hamel, P.B. and M.U. Chiltoskey. *Cherokee Plants: Their Uses – A 400 Year History*. Herald Publishing Company, 1975.

Hardy, B.L., et al. 'Direct Evidence of Neanderthal Fibre Technology and Its Cognitive and Behavioural Implications.' *Scientific Reports* 10, no. 1 (2020): 4889. https://doi.org/10.1038/s41598-020-61839-w.

Harris, S. 'Flax Fibre: Innovation and Change in the Early Neolithic: A Technological and Material Perspective.' In *Textile Society of America 2014 Biennial Symposium Proceedings: New Directions: Examining the Past, Creating the Future*, Los Angeles, California, 10–14 September 2014. University of Nebraska.

Haude, M.E. 'Identification of Colorants on Maps from the Early Colonial Period of New Spain (Mexico).' *Journal of the American Institute for Conservation* 37, no. 3 (1998): 240–70.

Hayase, S. 'American Colonial Policy and the Japanese Abaca Industry in Davao, 1898–1941.' *Philippine Studies* 33, no. 4 (1985): 505–17. http://www.jstor.org/stable/42633569.

Heath, C. 'El tiempo nos venció: Los Shipibo del Ucayali.' In *Una ventana hacia el infinito: Arte Shipibo–Conibo*, edited by P.P. Alayza and F. Torres. Instituto Cultural Peruano Norteamericano (ICPNA), 2002.

Hermkens, A-K. 'Barkcloth as Permeable and Perishable Substance in New Guinea Ontologies.' *Pacific Arts: New Series* 18/19 (2018–2019): 9–21.

Hernández, F. *Cuatro Libros de la Naturaleza y Virtudes de las Plantas y Animales*. Translated by F. Ximénez. Viuda de Diego López Dávalos, 1615.

Holland, J.H. 'Brazil-Wood.' *Bulletin of Miscellaneous Information, Royal Botanic Gardens, Kew*, no. 9 (1916): 209–25. https://doi.org/10.2307/4114329.

Von Humboldt, A. and Bonpland, A. *Personal Narrative of Travels to the Equinoctial Regions of the New Continent, During the Years 1799–1804*. Translated by H.M. Williams. Vol. IV. Longman, Hurst, Rees, Orme, and Brown, 1819.

Hurry, J.B. *The Woad Plant and Its Dye*. Oxford University Press, H. Milford, 1930. https://archive.org/details/woadplantitsdye0000unse.

Imbruglia, G. *The Jesuit Missions of Paraguay and a Cultural History of Utopia (1568–1789)*. Translated by M. Weyr. Brill, 2017.

Jolie, E.A., et al. 'Cordage, Textiles, and the Late Pleistocene Peopling of the Andes.' *Anthropology Faculty Publications* 136 (2011).

Keehnen, F.W.M. and A.A.A. Mol. 'The Roots of the Columbian Exchange: An Entanglement and Network Approach to Early Caribbean Encounter Transactions.' *Journal of Island and Coastal Archaeology* 16, no. 2–4 (2020): 261–89. https://doi.org/10.1080/15564894.2020.1775729.

Kennedy, J. 'Bananas and People in the Homeland of Genus Musa: Not Just Pretty Fruit.' *Ethnobotany Research and Applications* 7 (2009): 179–97.

Kriger, Colleen E. 'Mapping the History of Cotton Textile Production in Precolonial West Africa.' *African Economic History* 33 (2005): 87–116.

Kubley, A. and A. Damman *Tejiendo Quienes Somos (Weaving Who We Are): The Maya Youth Artisan Initiative*. Exhibition catalogue Mérida: Museo de Arte Popular Yucatán. University of Cincinnati Office of Research, 2021.

Kvavadze, E., et al. '30,000-Year-Old Wild Flax Fibers.' *Science* 325, no. 5946 (2009).

LaBerge, M. *The Heart of the Madder: An Important Prehistoric Pigment and Its Botanical and Cultural Roots*. Thesis. University of Wisconsin-Milwaukee, 2018.

Las Casas, B. de. *A Brief Account of the Destruction of the Indies*. Translated by F.A. MacNutt. Reprint of 1689 edition. Project Gutenberg, 2007. https://www.gutenberg.org/cache/epub/20321/pg20321-images.html.

Las Casas, B. de. *The Diario of Christopher Columbus's First Voyage to America, 1492–1493*. Translated by O. Dunn and J.E. Kelley Jr. University of Oklahoma Press, 1989.

Loftin, J.D. and B.E. Frey. *People of Kituwah: The Old Ways of the Eastern Cherokees*. University of California Press, 2024.

Loudon, N.N., M. Wollstonecroft and D.Q. Fuller. 'Plants to Textiles: Local Bast Fiber Textiles at Pre-Pottery Neolithic Çatalhöyük.' *Journal of Archaeological Science: Reports* 49 (2023).

Martínez García, M.J. 'Cheapening the Luxury: Some Curious Recipes with Vegetal Dyes.' In Luxury and Dress: Political Power and Appearance in the Roman Empire and Its Provinces, edited by C. Alfaro Giner, J. Ortiz García and M.J. Martínez García. Publicacions de la Universitat de València, 2013.

Medina, J.T. 'Carvajal's Account.' In *The Discovery of the Amazon According to the Account of Friar Gaspar de Carvajal and Other Documents*, edited by H.C. Heaton, compiled by J.T. Medina. Special Publication no. 17. American Geographical Society, 1934.

Melía, B. *El Guaraní: Experiencia Religiosa. Asunción: Biblioteca Paraguaya de Antropología*. In E. Dussel, *The Invention of the Americas: Eclipse of 'the Other' and the Myth of Modernity*, translated by M.D. Barber. Continuum, 1995.

Millar, G. *Orellana: Discoverer of the Amazon*. William Heinemann Ltd, 1954.

Neves, E.G., et al. 'Peoples of the Amazon before European Colonization.' In *Amazon Assessment Report 2021*, edited by C. Nobre, et al. United Nations Sustainable Development Solutions Network, 2021.

Parrish, J. 'On the Cultivation and Manufacture of Woad.' *The Annual Register, 1811*. Annual Register, 1811.

Pastoureau, M. *Blue: The History of a Color*. Princeton University Press, 2018.

Pliny the Elder. *The Natural History*. Translated by J. Bostock and H.T. Riley. Taylor and Francis, 1855, 17:17 and 35:27.

Polo, M. *The Travels of Marco Polo*. Translated by R. Latham. New York, 1957.

Prescott, W.H. *History of the Conquest of Peru*. Bickers & Son, 1890.

Radding, C. 'The Children of Mayahuel: Agaves, Human Cultures, and Desert Landscapes in Northern Mexico.' *Environmental History* 17, no. 1 (January 2012): 84–115. https://www.jstor.org/stable/23212617.

Ravenel, H.H. *Women of Colonial and Revolutionary Times: Eliza Pinckney*. Charles Scribner's Sons, 1896.

Reeves, W.P. *The Long White Cloud: 'Ao Tea Roa'*. 1899. eBook #12411. Project Gutenberg. Released May 1, 2004. https://www.gutenberg.org/files/12411/12411-h/12411-h.htm.

Ruiz, Walter. *Revaloración e Importancia del Aguaje*. Centro de Conservación, Investigación y Manejo de Áreas Naturales (CIMA), 2012.

Schick, T. 'Perishable Remains from the Nahal Hemar Cave: Precis of a Lecture Delivered at a Meeting of the Israel Prehistoric Society.' *Mitekufat Haeven: Journal of the Israel Prehistoric Society* (November 1985).

Schmidt, L.S. *American Involvement in the Filipino Resistance Movement on Mindanao During the Japanese Occupation, 1942–1945*. Master's thesis, U.S. Army Command and General Staff College, 1982. https://apps.dtic.mil/sti/pdfs/ADB068659.pdf.

Society for the Diffusion of Useful Knowledge. *Vegetable Substances: Materials of Manufactures*. Charles Knight, 1833.

Solazzo, C. and K.E.H. Penkman. 'Identification of the Earliest Collagen- and Plant-Based Coatings from Neolithic Artefacts (Nahal Hemar Cave, Israel).' *Scientific Reports* 6 (2016): 31053.

Splitstoser, J., et al. 'Early Pre-Hispanic Use of Indigo Blue in Peru.' *Science Advances* 2, no. 9 (2016).

Sztutman, R. and J.F. Sauma. 'When Metaphysical Words Blossom: Pierre and Hélène Clastres on Guarani Thought [Trans.].' *Common Knowledge* 23, no. 2 (2017): 325–44. Duke University Press.

The Gael: A Monthly Journal Devoted to the Preservation and Cultivation of the Irish Language and the Autonomy of the Irish Nation. 'Old Irish Costume and Customs.' December 1900, p. 351.

Theis, F.C. '22. Dyeing-process.' In *'Khaki' on Cotton and Other Textile Material*, 61–63. De Gruyter, 1903. https://doi.org/10.1515/9783112671023-024

'The Spanish Broom as a Fibre Plant (*Spartium junceum, L.*).' *Bulletin of Miscellaneous Information (Royal Botanic Gardens, Kew)* 1892, no. 63 (1892): 53–58. https://doi.org/10.2307/4102460.

Tió, F.E. *Los Conservadores Revolucionarios Yucatecos: Periodismo, Liderazgos y Prácticas de Prensa en la Construcción del Yucatán Revolucionario, 1897–1912*. Doctoral thesis, Centro de Investigaciones y Estudios Superiores en Antropología Social, 2016. http://ciesas.repositorioinstitucional.mx/jspui/handle/1015/1057.

Troldtoft Andresen, S. and S.Karg. 'Retting Pits for Textile Fibre Plants at Danish Prehistoric Sites Dated Between 800 B.C. and A.D. 1050.' *Vegetation History and Archaeobotany* 20, no. 6 (November 2011). HistoriskAtlas.dk.

Verschuuren, B., et al., eds. 'Cultural and Spiritual Significance of Nature: Guidance for Protected and Conserved Area Governance and Management.' *Best Practice Protected Areas Guidelines Series*, no. 32.

Zumbuhl, H. *'Tintes Naturales' (Natural Dyes)*. Huancayo, 1986.

# INDEX

Page numbers for glossary entries are in **bold**.

# Index

# Index

## ABOUT THE AUTHOR

A visionary changemaker, fashion designer, social entrepreneur and campaigner, Carry Somers is renowned for her transformative impact on the fashion industry. As co-founder of Fashion Revolution, the world's largest fashion activism movement, her work ignited a shift towards transparency, fairness and sustainability in supply chains.

With an MA in Native American studies, Carry has long supported Indigenous communities, amplifying their voices and embedding their values in her practice. She founded the award-winning Fair Trade brand Pachacuti in 1992 and co-founded League of Artisans, a non-profit championing artisanal skills, as a response to global challenges.

Recognised by Business of Fashion as one of the most influential figures shaping the industry, Carry appeared daily on the Nasdaq billboard in Times Square during New York Fashion Week 2022, which highlighted women-led social enterprises driving social and environmental impact in the fashion industry. Carry has also sailed two thousand miles researching ocean plastic pollution and collaborated on an award-winning textile garden at RHS Chelsea Flower Show. Her current roles include co-lead for Kew Gardens Community Open Week and adviser to Kew's Material World Festival in Autumn 2025.

She holds an honorary doctor of letters from Keele University and is an executive member of the International Academy of Digital Arts and Sciences. Awarded a Churchill Fellowship to carry out research into botanical fibres and dyes, Carry spent two months in Latin America in autumn 2023 meeting artisans, community activists and innovators, gathering rich material for the final part of this book.